THE BOOK OF INTELLIGENCE AND BRAIN DISORDER

Your Brain Must Have All Forms of Intelligence: IQ, EQ, and CQ

Amin Elsersawi, Ph.D. &

Dima Elsersawi, B.Sc. Psyc +GPS

TRAFFORD
PUBLISHING™

Order this book online at www.trafford.com
or email orders@trafford.com

Most Trafford titles are also available at major online book retailers.

The information, ideas, and suggestions in this book are not intended as a substitute
for professional medical advice. Before following any suggestions contained in this
book, you should consult your personal physician. Neither the author nor the publisher
shall be liable or responsible for any loss or damage allegedly arising as a consequence
of your use or application of any information or suggestions in this book.

Printed in the United States of America.

ISBN: 978-1-4269-4466-6 (sc)
ISBN: 978-1-4269-4467-3 (ebk)

*Our mission is to efficiently provide the world's finest, most comprehensive book publishing
service, enabling every author to experience success. To find out how to publish your book,
your way, and have it available worldwide, visit us online at www.trafford.com*

Trafford rev. 09/21/2010

 www.trafford.com

North America & international
toll-free: 1 888 232 4444 (USA & Canada)
phone: 250 383 6864 ♦ fax: 812 355 4082

5

Being gifted is something to be jealous about:

Being stupid is something to be kindhearted about:

Keywords

Genes, heredity, genetic regulation and crosses, structure of the brain, neurons and nerves, brain intelligence, change in the thickness of the brain's cortex, neurobiology of intelligence, white and gray matters, fluid and crystallized intelligence, classification and cognitive process, artificial neural networks, genotype intelligence, bodily intelligence, cognitive elite, Bell curve, eugenics IQ, IQ tests, method of IQ testing, factors affecting IQ tests, general ability index, Gifted children and individuals, environment and genetics contribute to giftedness, gifted drawers, savants, emotional intelligence (EQ), cultural intelligence(CQ), brain plasticity, brain training and fitness, executive brain, brain neoplasm and diseases, case study, and glossary.

Introduction

The scientific study of the brain and nervous system is called neuroscience or neurobiology. Because the field of neuroscience is so vast, and the brain and nervous system are so complex, this book will describe the structure and function of the brain and nervous system. You will be able to understand a number of tasks such as motor control, visual processing, auditory processing, sensation, learning, memory, and intelligence.

There are other introductions to human brain anatomy, but this is a book with a difference. It provides an easy and enjoyable means of learning and reviewing the fundamentals of the human brain through key concepts and progresses in small, logical, easy-to-learn increments. It is ideal for the nonexpert-students, professionals and amateur people alike.

This book explores the latest findings that demonstrate, through the use of technology like brain scans, that the old-aged brain is more flexible and more capable than previously thought. For the first time, long-term studies show that our view of the young-aged brain has better function than an old-aged brain which has been misleading and incomplete.

The book debunks three common myths about the brain and intelligence:

a. For many years, scientists thought that the human brain stopped growing after infancy and simply decayed over time and its dying cells led to memory slips, fuzzy logic, negative thinking, and even depression. A decade ago, many biologists and psychologists would have scoffed at that idea that the brain of adults loses its ability to generate new cells and rewire itself as we aged. Science and research prove that each time a new skill is learned, such as speaking a foreign language, interpersonal skills, dancing, or playing an instrument, the brain may grow to adapt and acquire new routes, even well into the senior years. This development in the brain is called "Plasticity".

b. Another thought is that if the memory starts to decline, there will be no improvement and there is not much one can do. Studies show that memory can be restored through consistent mental challenge by novel stimuli which increases production and interconnectivity of neurons and nerve growth factor, as well as prevents loss of connections and cell death. Researchers found that improvements in cognitive ability roughly counteract the degree of long-term cognitive decline typical among older people without dementia.

    c. Most of CEO's, legislators, judges, economists, doctors, lawyers, ministers and presidents are not in their 30s or 40s, but seasoned are veterans who bestow several decades of experience and expertise. Along with grey hair comes both knowledge and wisdom and you have to look far to find inspiring stories of accomplishment, creativity, and reinvention in the second half of life.

Dr. Elsersawi, one of the authors of this book, is an electrical engineer by training and a biochemist by inclination. He represents the brain as a closed box; of which the input can be one electrical pulse, and the output is of multi-order feed back loops. For example, you can measure the voltage, current, frequency of the electrical signals input to the brain, but you can not measure simultaneous individual signals distributed among the parts of the brain. Each individual signal passes though complicated pathways to execute certain functions such as perception, analysis, and task. The voltage is determined primarily by the potassium and sodium ionic concentrations internal and external to the neuron, about 70mv at rest. Current flow would be ionic not electronic and is not measurable.

Jeff Hawkins is an electrical engineer by training, and a neuroscientist by inclination. In his book (*How a New Understanding of the Brain will Lead to the Creation of Truly Intelligent Machines*) he used electrical engineering concepts as well as the studies of neuroscience to formulate his framework. In particular, Hawkins treats the propagation of nerve impulses in our nervous system as an encoding problem, specifically, a future predicting state machine, similar in principle to feed-forward error-correcting state machines.

Dr. Elsersawi believes that the efficiency of the internal workings of all individual signals inside the brain is called the intelligence. This is the overall picture of intelligence, yet we have no productive theories and facts about what intelligence is or how the brain works as a whole. The United States alone has thousands of neuroscientists. Most of them don't think much about overall theories of the brain because they're engrossed in doing experiments to collect more data about the brains many parts. And although legions of computer programmers have tried to make computers intelligent, they have failed. Many neuroscientists tend to reject or ignore the idea of considering the brain in computational terms, and computer scientists often don't believe they have anything to learn from biology.

The book starts with details on the structure of the brain and how the physical brain implements the memory-prediction model, i.e., how the brain actually works.

Chapter 2 of this book describes what intelligence is and how your brain deals with it. This chapter is overflowing with information, and is a great starting point for venturing through the neuropsychology world of intelligence. It discusses sensitive topics such as interpersonal and intrapersonal skills and related concepts, such as social skills, psychological maturity and emotional awareness.

It also provides a handful of sample questions on IQ and EQ. Such questions offer readers the side benefits of finding out if they themselves possess a high level of intelligence.

Other interested topics including cognitive processes, intelligent theories, environmental and genetic contribution to giftedness, autism, cultural intelligence, brain plasticity and brain trust programs are also discussed in chapter 2. Chapter 3 includes lists of brain diseases and disorders. Topics discussed include: neurotransmitter disorders, inborn errors, brain fluids, demyelination, brain cancer, brain trauma, Alzheimer's, Parkinson's, and even schizophrenia.

The definitions of terms given within the text are supplemented by a comprehensive glossary, and there are many informative images.

At the end of the book, the reader is guided through a case study that illustrates the philosophy of cognitive abilities in children.

Readers will find this book an inspiration-provoking read and an impetus for new theories of embodied cognitive science. It will be particularly helpful for people interested in getting involved in the understanding of intelligent agents.

Chapter 1

Genes and Heredity

1. Cell

The cell is the smallest structural and functional unit of all living organisms. It is often called the building block of life. Some organisms are unicellular like ameba and bacteria, and some are multicellular like humans. Humans have about 100 thousand billions of cells with a diameter of about 10 micrometers and a weight of 1 nanogram, Figure (1.1).

Figure (1.1): One cell with the cytoplasm and nucleus

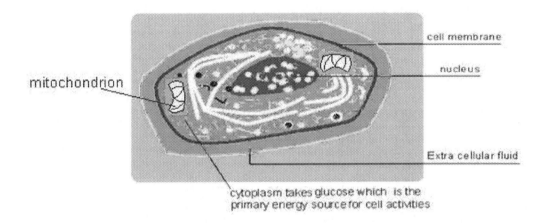

The Cytoplasm needs energy that is taken from glucose. Glucose and fructose have a similar number of carbon, hydrogen, and oxygen atoms. Both have the formula $C_6H_{12}O_6$. The difference is that glucose is hexagonal and has only one double bonded oxygen, and fructose is pentagonal and has two double bonds of oxygen. Both are isomers and interchangeable under the process of photosynthesis. In addition to the carbohydrates, lipids, and proteins, the cell needs something called nucleic acid. Carbohydrates have a unique formula which is the ratio between carbon, hydrogen, and oxygen which is about 1:2:1. Lipids, protein, and nucleic acids have different ratios.

Nucleic acid is used by the cell to store and use hereditary information. Nucleic acids known as nucleotides consist of three main components: a nitrogenous group, a phosphate group, and a sugar.

There are two types of nucleotides distinguished by their sugar: Deoxyribonucleic acid (DNA), which has one less oxygen than the other type, ribonucleic acid (RNA). The function of DNA molecules is to store genetic information for the cell. RNA molecules carry genetic messages from the DNA in the nucleus to the cytoplasm for use in protein synthesis and other processes. We shall discuss nucleic acids in detail in the coming sections.

1.1 Mitosis and Meiosis

1.1.1 Mitosis

Mitosis is the process by which a eukaryotic cell separates the chromosomes in its cell nucleus into two identical sets in two nuclei, [http://en.wikipedia.org/wiki/Mitosis]. It is generally followed immediately by cytokinesis, which divides the nuclei, cytoplasm, organelles and cell membrane into two cells containing roughly equal shares of these cellular components. Mitosis has two daughter cells, somatic cells (body cells), it repairs the body cells, identical cells, diploid and it takes 5 phases. Mitosis and cytokinesis together define the mitotic (M) phase of the cell cycle (see cell cycle) - the division of the mother cell into two daughter cells, genetically identical to each other and to their parent cell. This accounts for approximately 10% of the cell cycle.

Eukaryotic cells reproduce in two different ways: mitosis and meiosis. In mitosis, the nucleus divides into two nuclei (referred to as karyokinesis), and the cytoplasm divides into two portions (referred to as Cytokinesis), Figure (1.2).

Figure (1.2): Reproduction of two cells from one cell (mitosis)

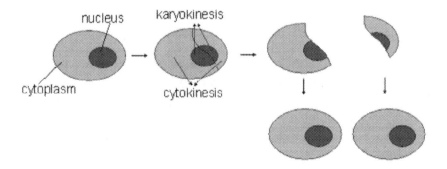

In the karyokinesis division, the chromosomes separate first and then are followed immediately by cytokinosis, in which the cell divides in to two identical daughters, Figure (1.3).

Figure (1.3): Division of nucleus (mitosis)

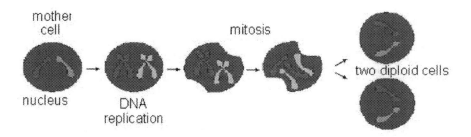

The period between successive divisions is known as interphase. The interphase period is divided into a number of stages, Figure (1.4).

Figure (1.4): Stages of cell division (mitosis)

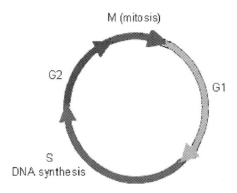

Some cells divide rapidly. For example, beans take about 19 hrs to complete the cycle division, whereas red blood cells divide at a rate of 2.5 million per second. Cancer cells divide rapidly. The daughter cells divide before they have reached maturity. Electrocharge, pH, temperature, and some drugs (enzymes) may affect the rate of division. When cells stop dividing, they stop at a point late in the stage G1. The stage S is the stage when the DNA is replicated for the next division, and the chromosomes become double stranded. The cell has then entered the G2 stage and proceeds in to cell division. Cells will not divide again and stop in the G1 stage. The process of division passes into the following steps:

1.1.1.1 Prophase

Prophase is the first step of mitosis preparation where the cell is about to divide, and the chromosomes become visible and start to condense into double stranded chromosomes, Figure (1.5). Chromatin/DNA do not replicate in this step. Gradually, a spindle composed of protein fibers together with kinetochores forms and extends nearly the length of the cell, expanded in its centre (or the equator of the cell) like a base ball. The chromosomes each consist of two chromotids attached to each other by a spindle fiber at the centromere.

Figure (1.5): Double stranded chromosome divides into two identical sisters

1.1.1.2 Metaphase

When the centromere and kinetochore arrive at the centre of the cell (equator), the metaphase begins. The mitosis process ends when the centromere divides so that each of the chromatids becomes a single stranded chromosome, Figure (1.6).

1.1.1.3 Anaphase

During anaphase, each sister of a single-stranded chromosome moves towards one pole of the cell (one sister to one pole, and the other to the opposite pole). Thus, anaphase is the stage when sisters of the chromosome migrate to the two poles of the cell, Figure (1.6).

Figure (1.6): Movement of daughter chromosomes from the equator to poles during anaphase

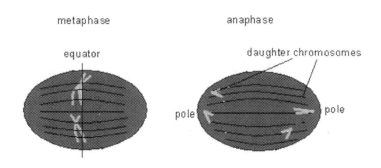

1.1.1.4 Telophase

In telophase, the nuclear envelope (which had disappeared during prophase) reforms, and chromosomes uncoil into the chromatin form again. There are now two cells instead of one cell, but they are of smaller sizes of the same genetic heredity. The small cells will develop into mature ones.

1.1.1.5 Cytokinesis

Cytokinesis is the last stage of cell division, where the daughter cells split apart. The mitosis is the division of the nucleus, and the cytokinesis is the division of the cytoplasm.

1.1.2 Meiosis

Meiosis (sex cells) is genetically different from mitosis. The process of cell division in sexually reproducing organisms that reduces the number of chromosomes in reproductive cells from diploid to haploid, leading to the production of gametes in animals and spores in plants. In the meiosis process there are four cells (sperm-gametes and eggs) cross over (haploid), and it takes 9 phases.

Meiosis produces daughter cells that have one half the numbers of chromosomes as the parent cell.

During meiosis chromosomes are duplicated once in S stage. Meiosis has two stages; the meiosis 1 is just like mitosis where the cell divides once. In meiosis 2 the cell divides again and the final product is four daughter cells. The process of meiosis is as follows:

- o The first step in meiosis is the diploids of each of the 46 chromosomes duplicated, Figure (1.7). This is called DNA replication.
- o The homologues of chromosomes (23 homologues in 46 chromosomes; i.e. one chromosome has two homologues) are segregated in two cells.
- o Homologues pair up alongside, lengthwise, no.4 of Figure (1.7).
- o The pairs of homologues are shuffling and cross over in metaphase 1, Figure (1.7).
- o At the end of meiosis, new chromosomes will be created to become part of eggs and sperms, which will be unique to their parent.
- o During meiosis 2, the sister chromatids of each of the 23 chromosomes are pulled apart resulting in 4 cells. Each of the 23 chromosomes is now haploid.
- o The Meiosis process is the way to create the diversity of all sexually reproducing organisms.

Figure (1.7): Division of nucleus (meiosis)

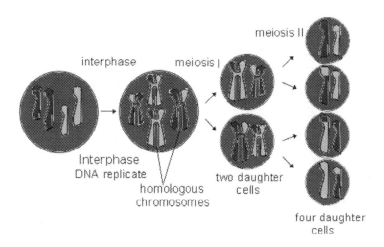

The invention of electronic microscopes allowed biologists to discover the basic facts of cell division and sexual reproduction. The centre of genetics and heredity research then shifted to understanding what really happens in the transmission of hereditary traits from parents to children. A number of hypotheses were suggested by researchers but Gregor Mendel, a monk from Austria was the father of heredity.

1.2 Mitochondria

Mitochondria are central components of our cells that generate the majority of our energy from nutrients. However, their weak side is that, through their normal activity, they generate unstable chemicals that harm both the mitochondrion itself and other components of the cell. This resulting damage is thought to play an important role in aging.

Mitochondria are called the powerhouses of the cell. They are membrane enclosed organelles found in most eukaryotic cells (eukaryotic cells are in all animals except bacteria and cyanobacteria). They absorb the nutrients, break them down and then release energy called ATP, for the cell. The process of releasing energy is known as cellular respiration, which is associated with chemical reactions needed for the growth of the cell. Mitachondria are very small organelles. Some cells have one mitochondrion and others have several thousand mitochondria. Cells for transmitting nerve impulses have fewer mitochondria than muscle cells that need high energy. The numbers of mitochondria are adjusted depending on the energy needed by the cell. Mitochondria can move, grow, and combine with other mitochondria inside the cell. Figure (1.8) is of a mitochondrion with the outer and inner membranes. The purpose of Cristae is to increase the surface area inside the cell for more chemical reaction, when required. The matrix inside the mitochondria is fluid. Inside the matrix, oxygen is combined with protein to digest the food molecules, and the water inside the matrix, takes the digested food to feed the cell.

Mitochondria are the only places in the cell where oxygen is used to digest the food molecules. Calcium may also be controlled by mitochondria.

Mitochondria have their own independent genome, the material of which is known as mitochondrial DNA (mtDNA).

Figure (1.8): Human mitochondrion with its DNA, cristae, matrix, and ribosome

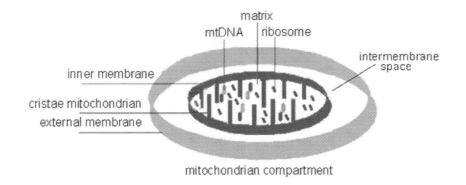

One cell may have several thousand mitochondria

1.3 Chromosomes

A chromosome is an organized structure of DNA and protein that is found in the nucleus of each cell. It is a single piece of coiled DNA containing many genes, regulatory elements and other nucleotide sequences. In humans, each cell normally contains 23 pairs of chromosomes, for a total of 46. Twenty-two of these pairs, called autosomes, look the same in both males and females. The 23rd pair, the sex chromosomes, differs between males and females. Females have two copies of the X chromosome, while males have one X and one Y chromosome, Figure (1.9). Chromosomes also contain DNA-bound proteins, which serve to package the DNA and control its functions.

Figure (1.9): Chromosomes

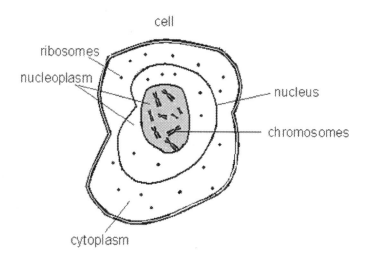

1.4 Cell Cycle

The division of the cell happens in one cycle of the cell cycle. In normal division, both the centrosome and the chromosomal DNA must be duplicated only once per one cycle of the cell cycle. One cell cycle passes through four phases: G1, S G2, and M. In the G1 phase, there is one centrosome per cell that consists of two perpendicular centrioles. The duplication process of the centrosome starts at the phase G1-S transition, at which the DNA also starts replication. In this phase (G1-S), the two perpendicular centrioles start to separate from each other. Once separated, centrioles start to grow orthogonally. At G2, each centriole becomes a pair of centrioles, Figure (1.10). This pair of centrioles has one new and one old centriole. In the duplication of somatic cells, there must be an old centriole to create a new centriole. This is not the case in all animals and plants where new centrioles are not of the same pattern.

Centrosomes are the microtubule-organizing centers and are important for the formation of the bipolar spindle during mitosis. The centrioles are surrounded by a poorly defined electron-dense protein matrix called the pericentriolar material (PCM). Aberrant centrosome numbers can result in the generation of multipolar and monopolar spindles, potentially causing cell death.

During the centrosome cycle the key enzymes of the cell cycle is regulated by Cdc-25 phospatase which regulates the enzyme of the cell cycle, the cyclin-dependant kinase (Cdk) during the phase G1S and the phase G2M.

Figure (1.10): Nucleus and centrosome cycle

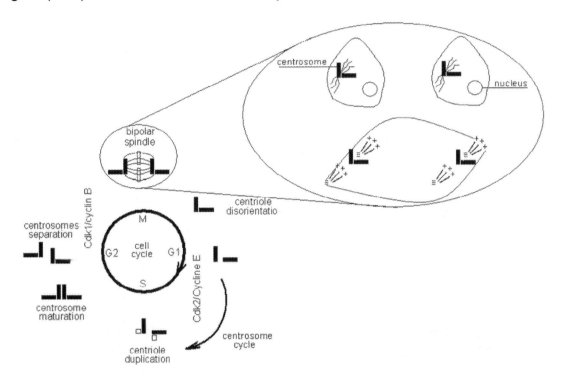

At the G2-M transition phase the new centrosomes meet the chromosomes and interact with them to create the bipolar spindle. This causes the chromosomes to separate. Recent research has shown that cyclin E and its associated E-Cdk2 are required for centrosome duplication, and this reaches its peak of activity at the G1-S transition. Also, the initiation of the DNA replication takes place at this stage.

1.5 Cell Respiration

The process of respiration serves two basic purposes in living organisms, the first one being the disposal of electrons generated during catabolism and the second one being production of ATP during anabolism. The respiration machinery is located in cell membranes of eukaryotes, whereas it is placed in the inner membranes of the mitochondria. Respiration requires a terminal electron acceptor which uses oxygen as its terminal electron acceptor, and it is called aerobic respiration. The one, which uses terminal electron acceptors other than oxygen, is called anaerobic respiration.

There are two types of respiration: aerobic and anaerobic.

1.5.1 Aerobic Respiration

Aerobic respiration is the process that takes place in presence of oxygen which is used as the terminal electron acceptor. Fuel molecules commonly used by cells in aerobic respiration are glucose, protein and fat. The process of obtaining energy in aerobic respiration can be represented in the following equation:

Glucose + Oxygen → Energy + Carbon dioxide + Water

$$C_6H_{12}O_6 + 6O_2 \rightarrow ATP + 6CO_2 + 6H_2O$$

Aerobic cellular respiration has four stages. Each is important, but could not happen without those preceding it. The steps of cellular respiration are:

1.5.1.1 Glycolysis (the break down of glucose), Figure (1.11).

Glycolysis involves the breakdown of glucose. Cells obtain glucose from the blood. Blood glucose levels are maintained by the interaction of two processes: glycogenesis and glycogenolysis. Glycogenesis is the production of glycogen from glucose and occurs primarily in the liver and skeletal muscles. This occurs when blood glucose levels are too high, for example, after a meal. When blood glucose levels begin to decline, the glycogen starts to breakdown to release molecules of glucose in a process called glycogenolysis. This occurs during hunger and several hours after a meal.

The breakdown of glucose releases hydrogen which will be used to phosphorylate (add phosphor) the ADP to ATP. The release of hydrogen will cause reduction (opposite to oxidation) to the ADP, i.e., the ADP becomes more negative (it gains electrons). The phosphate atom, which ends with 5 electrons, will bind to the ADP, forming ATP (energy). The following equation represents the process of the glycolysis:

Glucose → 2 Pyruvic Acid (or pyruvate) + 2 net ATP + 4 hydrogens (2 NADH2)

Glucose + 2ATP → Glucose + 2P + 2ADP
Glucose + 2P + 2ADP → 2 (Pyruvate + P) + 2NAD⁺ + 4 Hydrogen
2 (Pyruvate + P) + 2NAD⁺ + 4 Hydrogen → 2 (Pyruvate + P) + 2NADH + 2
Hydrogen → 2 Pyruvate +2P +2 ADP + 2 Nicotinamide + 2 Hydrogen → 2
Pyruvate + 2 ATP +2 ATP

The following diagram represents the hydrogen transfer in which pairs of hydrogen are successively passed from one substance to another, and these substances are called hydrogen carriers.

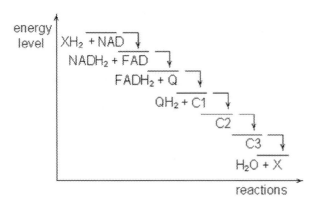

Figure (1.11): The transformation of glucose to acetyl coenzyme CoA

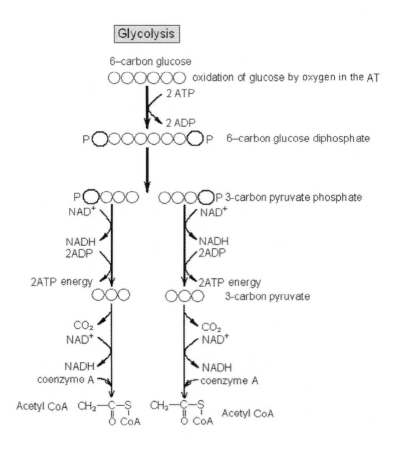

1.5.1.2 Pyruvate Decarboxylation

Pyruvate decarboxylation is when pyruvate (through glycolysis) has carbon dioxide removed from it (called decarboxylation) and hydrogen, Figure (1.12).

Figure (1.12): Pyruvate decarboxylation

The first part of the decarboxylation which includes CoA-SH will go to the Krebs cycle.

The coenzyme A consists of three parts:

a. The adenosine diphosphate (ADP)

b. A protein

c. A sulph-hydryl (-SH)

Pyruvate is also dehydrogenated by the enzyme dehydrogenase. So, the pyruvate dehydrogenates by the enzyme dehydrogenase, and is decarboxylized by the enzyme CoA. The release hydrogen is accepted by NAD^+ (oxidation), and the product of oxidative decarboxylation is an acetyle group, which is accepted by CoA to make acetyle-CoA. This process happens in the mitochondria.

1.5.1.3 Krebs Cycle

During the Krebs cycle, the following steps happen:

1. There are three interlinked energy production cycles:
    o the glycolitic (sugar burning)
    o the Krebs' citric acid cycle (protein and fat burnings)
    o electron transport

2. At the end of the glycolitic cycle (pyruvic acid), electrons spark helping power the Krebs cycle to generate much of the ATP

3. At the beginning of the Krebs cycle, protein and fat need oxygen to burn them. This oxidation process is accomplished by the NAD which takes the hydrogen from foods to ease the oxidation. The NAD become oxidized, and the NADH is reduced (electron-energy rich).

4. The NADH is rich with electrons, and therefore generates much of the ATP energy that literally energizes our life. Each unit of NADH is capable of generating three units of ATP energy. When there is a lot of energy (ATP) in

23

the body, it means that the food has lots of oxygen (or the food has been well oxidized). The ATP is generated due to the oxidative break down of pyruvate.

The question is "can ATP be produced by the non-oxidative breakdown of the pyruvate"? This is the cause of CANCER as stipulated by Warburg's hypothesis which he won the noble prize for. Otto Warburg (1883) hypothesized that cancer cells generate energy (ATP) by the non-oxidative breakdown of pyruvate (in the process that is called glycolisis). This is in contrast to "healthy" cells which mainly generate energy from oxidative breakdown of pyruvate. Pyruvate is an end-product of glycolysis, and is oxidized within the mitochondria. Therefore, according to Warburg, cancer should be interpreted as the dysfunction of the mitochondria. Warburg reported a fundamental difference between normal and cancerous cells to be the ratio of glycolysis to respiration; this observation is also known as the "Warburg effect". He quoted "the prime cause of cancer is the replacement of the respiration of oxygen in normal body cells by a fermentation of sugar". In fermentation the ATP produced is three units, whereas the ATP produced through the glycolisis process of normal cells is 36 units. Figure (1.13) shows the Krebs cycle.

Figure (1.13): Krebs cycle

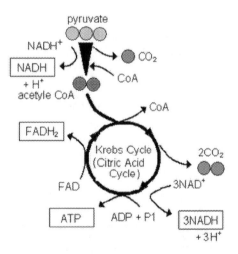

The overall ATP production from glucose is 38 distributed as follows:

8 ATPs from glycolysis of the glucose to pyruvate

6 ATPs from the hydrogens produced when the two pyruvic acids are converted into two acetyl CoAs

24 ATPs from the Krebs cycle

Proteins and lipids are also used as a source of energy.

Proteins are first broken down into amino acids. The nitrogen component of amino acids is then removed (in a reaction called deamination), and these deaminated amino acids are then converted into Acetyl CoA which passes through the Kreb's Cycle to make more ATP.

Lipids (Fatty Acids) are converted into molecules of Acetyl CoA in a process called Beta oxidation. Fatty acids are oxidized by most of the tissues in the body. However, the brain, erythrocytes, and adrenal medulla cannot utilize fatty acids for energy requirements. Beta oxidation splits long carbon chains of the fatty acid into acetyl CoA, which can eventually enter the Krebs cycle. The splitting is performed by the oxidative degradation of saturated fatty acids in which two-carbon units are sequentially removed from the molecule with each turn of the cycle.

1.5.1.4 Electron Transport Chain

The 2,3-Bisphosphoglycerate is a three-carbon isomer of the glycolytic and is present in human red blood cells at approximately 5 mmol/L. It binds with greater affinity to deoxygenated hemoglobin at respiring tissues than it does to oxygenated hemoglobin (e.g., in the lungs). This Bisphosphoglycerate is the product of 2,3 bisphosphate binds with glyceraldehyde that is produced from the break down of the glucose (pyruvate), Figure (1.14).

Figure (1.14): 2,3-Bisphosphoglycerate

2,3-Bisphosphate   glyceraldehyde   2,3-Bisphosphoglycerate

The 2,3-Bisphosphoglycerate is reduced by two nucleotides: the adenine and the nicotinamide ribose. The product is the enzyme NAD^+, Figure (1.15).

The process of glycolysis produces two pyruvates or one glyceraldehyde bound to a bisphosphate or a monoposphate.

Figure (1.15): NAD$^+$

NAD$^+$ (Nicotinamide Adenine Dinucleotide)

The reduced compounds, such as protein, fat and glucose, are oxidized, thereby releasing hydrogen, as seen form the following equation (let's take the glucose, for example):

Glucose → Pyruvate + Hydrogen

$C_6H_{12}O_6$ → $C_3H_4O_3$ + $2H_2$

This hydrogen is transferred to NAD$^+$ by reduction to NADH, as part of glycolysis and the citric acid cycle. The NADH takes the proton of the hydrogen (H$^+$) and leaves the electron in the cytoplasm which is then transferred into the mitochondrion in a process called Electron Transport Chain (ETC). The Glycolysis process produces 2NADH, 2H$^+$, 2e$^-$, and 2 pyruvate molecules from one glucose molecule. The reduction of NADH is shown in Figure (1.16).

Figure (1.16): Reduction of NADH by H^+

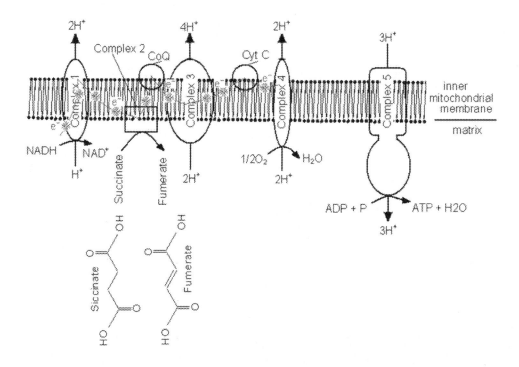

$$NAD^+ + H^+ + 2e^- \rightleftharpoons NADH$$

$$NAD^+ + 2H^+ + 2e^- \rightleftharpoons NADH + H^+$$

The electron transport chain is explained in the following steps, Figure (1.17).

Figure (1.17): Electron transport chain through the inner mitochondrial membrane

1- The electron starts when NADH loses its hydrogen, i.e. when the NADH is oxidized. Losing hydrogen is oxidation and gaining hydrogen is reduction.

2- The hydrogen's electron is passed through the first carrier protein in complex 1. The protein pumps the hydrogen's proton outside the membrane and passes the hydrogen's electron to other carrier proteins in the membrane, which is the complex 2.

3- Succinate combines with FAD (flavin adenine dinucleotide) which is a redox factor and available in the CoQ, Figure (1.18). The product will be Fumerate and FADH as per the following equation:

Succinate + FAD → Fumerate + FADH

Figure (1.18): Chemical reaction of succinate and fumerate

4- The falling electrons pump more protons (H^+) out.

5- The wandering electrons combine with oxygen to make water at complex 4 as per the following equation:

$$O_2 + 4H^+ + 4e^- \rightarrow H_2O$$

The accumulated protons (H^+) outside the membrane create a gradient of protons (a huge number of protons). The crowded number of protons returns back to the cytoplasm (matrix) of the mitochondrion through a special channel (complex 5). Each proton passing to the matrix provides enough energy to make one ATP out of one ADP as seen in the following equation:

28

ADP + Pi + H$^+$ $\rightarrow$ ATP + H$_2$O, ΔE = 30.54 kJ/mol (7.3kcal/mol)
ATP + H$_2$O $\rightarrow$ ADP + Pi + H$^+$, ΔE = -30.54 kJ/mol (-7.3kcal/mol)

Basically, ADP and ATP go in a cycle. ATP is like a rechargable battery that needs to be recharged. Oxygen burns the food and the energy is stored as ATP. When the organism needs energy, it uses that ATP as its energy source, and the ATP converts to ADP.

1.5.2 Anaerobic Respiration (Fermentation)

Fermentation is an alternative to cell respiration when cellular respiration is not possible. For example, organisms living in anaerobic conditions use the fermentation process. In the fermentation process, only glycolysis process happens, and is the first and last step of the cellular respiration process.

Glycolysis proceeds normally, as in aerobic conditions, producing a net gain of 2 ATP. The two pyruvate molecules, however, are reduced and the NAD necessary for the initiation of glycolysis is recycled. The cells do not deplete their store of NAD, although they are only able to produce 2 ATP. As a by-product of fermentation, either ethanol or lactic acid is produced.

1.6 DNA

DNA (DeoxyribonNucleicAcid) is a nucleic acid that carries the genetic information in the cell and is capable of self-replication, synthesis of RNA and translation of protein. DNA is located in the nucleus of all cells; it is the substance of the chromosome*s* that separate from the nucleus when cells divide, and it carries the genes. DNA consists of two long chains of nucleotides twisted into a double helix and joined by hydrogen bonds between the complementary bases (nucleotides) adenine and thymine or cytosine and guanine. The sequence of nucleotides determines individual hereditary characteristics by controlling protein synthesis in cells. The bases are adenine, cytosine, guanine, and thymine.

When the cell divides, its DNA also replicates in such a way that each of the two daughter molecules is identical to the parent molecule. The hydrogen bonds between the complementary bases on the two strands of the parent molecule break and the strands unwind. Using as building bricks, nucleotides present in the nucleus, each strand directs the synthesis of a new one complementary to itself. Replication is initiated, controlled, and stopped by means of polymerase enzymes.

1.6.1 Nucleic Acid and Genes

Nucleic acids (DNA and RNA) specify and aid the construction of proteins which provide the functional elements of biological systems. DNA stands for deoxyribonucleic acid and RNA stands for ribonucleic acid, both of which are composed of nucleotides, Figure (1.19).

Figure (1.19): DNA and RNA are composed of sugar, phosphate and nucleotides

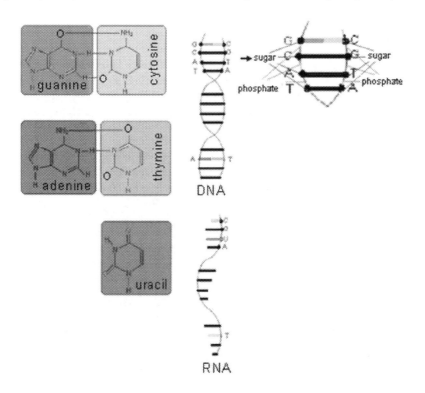

In our body, there are hundreds of thousands of proteins. These proteins are made locally. An enormous amount of information is required to manufacture them according to specific codes transmitted from the DNA to the exact location by the RNA.

The information required for designing each protein is stored in a set of molecules called nucleic acids. There are two types of nucleic acids: DNA and RNA. Both of them are made up of very large molecules that have two main parts. The nucleic acid has the backbone, which is made of alternating sugar and phosphate molecules bonded together in a long chain, as seen in Figure (1.19). Sugars in the backbones are connected to nucleotide bases called adenine (A), cytosine (C), guanine (G), and thymine (T). The DNA has two backbones, connected together by two nucleotide bases such that C and G is one bond, and A and T is another. Bonds are connected in alternate ways. Each nucleic acid contains millions of nucleotide bases. They are arranged in different groups of bonds so that each group represents a certain code of information carried to each specific protein with the same information as the genetic traits of parents. Each group is an exact genetic alphabet on which each of our proteins is coded. The two strands of the

DNA wrap around each other forming a coil, or helix. The DNA has the ability to copy itself and self-replicate when needed during cell division, cell growth, and DNA repair.

Copying or replication can be achieved in two steps. In the first step the hydrogen bonds between the nucleotide bases break and the two strands separate. In the second, new complementary bases are brought and paired up with the same bases of the DNA, thus forming identical DNA.

Ribonucleic acid (RNA) is similar to DNA in construction, except that RNA does not have thymine. Instead it has uracil (U), and has only one single strand of backbone with a similar construction to DNA (sugar and phosphate). RNA is used as a transporter and messenger of information from the DNA out of the cell to help manufacture protein.

There are mainly three types of RNA: messenger RNA (mRNA), transfer RNA (tRNA), and ribosomal RNA (rRNA). Each one of them has a special function. mRNA contains information on the sequence of amino acids stored in the DNA, and carries the genetic code from the DNA to the ribosomes. The rRNA is in the cytoplasm and combines with protein to make the ribisomes. The tRNA bonds to amino acids and then synthesizes the protein, Figure (1.20).

Figure (1.20): mRNA takes information on amino acids and equips the ribosome to manufacture the specific protein when it receives the amino acids that are delivered by tRNA.

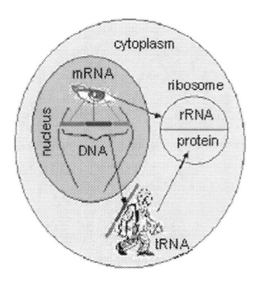

Transfer RNA (tRNA) is the information adapter (like electrical connector) molecule. It is the direct interface between an amino-acid sequence of a protein and the information in DNA.

Messenger or mRNA is a copy of the information carried by a gene on the DNA. The role of mRNA is to move the information contained in DNA to the translation machinery (tRNA).

Ribosomal RNA (rRNA) is a component of the ribosomes (sugar) and the protein synthetic factories in the cell.

The double helix of the DNA in detail is shown in Figure (1.21). The figure shows that the backbones are composed of sugar and phosphate. The two backbones look like a spiral stair case, and connect to bases of adenine (A), thymine (T), cytosine (C) and guanine (G). A and T are connected by two hydrogen atoms, and G and C by three hydrogen atoms.

Figure (1.21): DNA molecule with two views

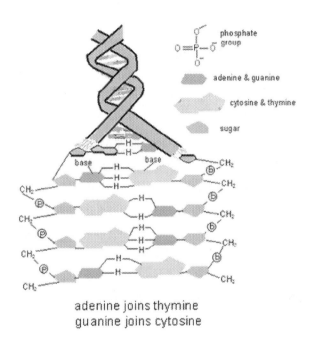

1.6.2 The Structure of Chromosomes, DNA and Cells

Cells can be different from each other because each cell or group of cells are specialized in a particular function such as sensing light (eyes' cells), fighting disease (immune cells), carrying oxygen (red blood cells), hair color, etc. One cell has a nucleus in the centre that has 46 chromosomes. The 46 chromosomes have more that 35,000 genes. Each chromosome is packed with DNA, genes, proteins, and other kinds of molecules, Figure (1.22).

32

Figure (1.22) Chromosome, DNA, and histones

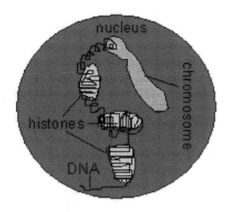

The Nobel Prize in Physiology or Medicine was awarded this year (2009) to three scientists who have solved a major problem in biology: how the chromosomes can be copied in a complete way during cell division and how they are protected against degradation. The Nobel Laureates have shown that the solution is to be found in the ends of the chromosomes – the telomeres – and in an enzyme that forms them – telomerase. Telomeres are in the spotlight of modern biology. Whether the subject at hand is cancer, gene regulation, organismal aging, or the cloning of mammals, much seems to depend on what happens at the ends of chromosomes.

Before we discuss the subject in detail, we should know the structure of chromosomes, DNA, and the cell, Figure (1.23).

Figure (1.23): The structure of chromosomes, DNA and cell

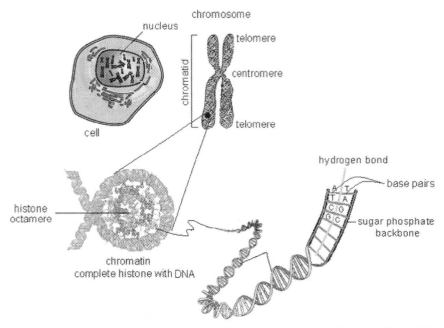

chromosome

nucleus

telomere

chromatid

centromere

telomere

cell

hydrogen bond

base pairs

histone
octamere

sugar phosphate
backbone

chromatin
complete histone with DNA

histone octamer consists of H3-H4 dimer (H3 + H4) andH2A - H2B dimer (H2A + H2A)

Chromosomes end with two telomeres, Figure (1.23). If the telomeres are shortened, cells and humans age. Conversely, if telomerase activity is high, or if the telomeres are extended, chromosomes (and DNA) are maintained, and cellular senescence is delayed. In contrast, if telomeres are damaged or defected, cancer and certain inherited diseases may be developed. Scientists found that when a cell is about to divide, the DNA molecules, which contain the four bases that form the genetic code, are copied, base by base, by DNA polymerase enzymes. However, for one of the two DNA strands, a problem exists in that the very end of the strand cannot be copied, therefore, the chromosomes should be shortened every time a cell divides, Figure (1.24).

Figure (1.24): Shortened DNA and chromosome due to a damaged or shortened telomere

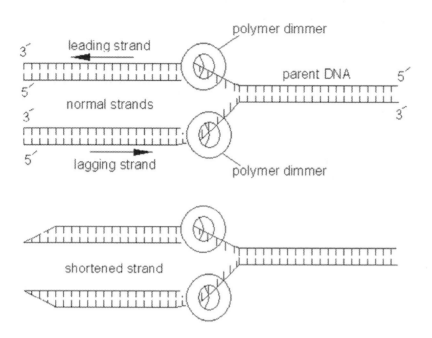

Luckily, these problems were solved when this year's Nobel Laureates (Elizabeth Blackburn, Jack Szostak, and Carol Greider) discovered how the telomere functions and found the enzyme that copies it.

1.7 Heredity in the Genes

DNA (deoxyribonucleic acid) is a nucleic acid that carries genetic information. The study of DNA launched the study of structural and functional properties of biological systems, altered genetic expressions, and led to the cracking of the biochemical code of life. Understanding DNA has facilitated genetic engineering and DNA testing which has led to:
  - involvement in the production and use of recombinant DNA
  - creation of bacteria that synthesizes insulin and other human proteins
  - genetic manipulation of various organisms which has enabled cloning
  - asexual reproduction of identical copies of genes and organisms
  - identification of individuals by the genetic fingerprinting
  - prediction, diagnosis, prevention and treatment of diseases
  - Determination if two people are related
  - Determination if two people descend from the same ancestors
  - Finding out if you are related to others with the same surname
  - Proving or disproving of your family tree research
  - Providing clues about your ethnic origin

Before we understand the heredity DNA and the functional properties of genetic engineering, we must discuss the actual journey into DNA.

1.7.1 Gregory Mendel – the Father of Genetics

Gregor Johann Mendel (July 20, 1822 – January 6, 1884) was an Augustinian priest and scientist, who gained posthumous fame as the figurehead of the new science of genetics for his study of the inheritance of certain traits in pea plants. He was an Augustinian monk who taught natural science to high school students. Mendel showed that the inheritance of these traits follows particular laws, which were later named after him. The significance of Mendel's work was not recognized until the turn of the 20th century. Mendel's work became the foundation for modern genetics.

Mendel derived certain basic laws of heredity: hereditary factors do not combine, but are passed intact; each member of the parental generation transmits only half of its hereditary factors to each offspring (with certain factors "dominant" over others); and different offspring of the same parents receive different sets of hereditary factors.

Mendel's experiment was based on the following combination, Figure (1.25).

Figure (1.25): Mendel's experiment

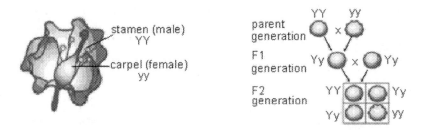

The name chromosome comes from the Greek word (chromo) meaning color, and (soma) meaning body. The word chromosome means uniqueness of a human. Chromosomes are thread-like structures located inside the nucleus of animal and plant cells. Each chromosome is made of DNA (deoxyribonucleic acid) that passed from parents to their offspring. Chromosomes vary in number and shape among living things. Humans have 46 chromosomes (23 pairs). Most bacteria have one or two circular chromosomes (a fruit fly has four pairs, and a dog has 39). Some offsprings inherit their traits from their mothers and others from their father. Only mothers keep chromosomes of their mitochondria during fertilization of egg cells. Females have two X chromosomes in their cells, and males have one X and one Y chromosome. Inheriting imbalance of chromosomes may lead to health problems. For example, women with only one X chromosome are short and have some heart and kidney problems. Men who have more X chromosomes than Y have tall stature.

One cell has twenty two autosomes (chromosomes divide through the mitosis process and are the same in both sexes in humans, and have been used in genetic testing for ancestry, ethnicity, race, and genealogical purposes), and one sex chromosome. The longest chromosome is referred as chromosome 1, the next longest as chromosome 2, and so on to the smallest autosome, chromosome 22.

In his experiments on peas plants Mendel considered traits such as colour, height, shape and other visible traits. He did not account for hidden characters and traits such as taste, emotion, intelligence, age, and other factors dominating animal behavior. Offspring may or may not be identical physical copies of parents; rather they are the combination of all character traits, whether visible or invisible, sensible or insensible. Let's take, for example the first law of Mendel, Table (1.1).

Table (1.1): Genotypes crossing

	Y	y
Y	YY	Yy
y	Yy	yy

Three genotypes have traits influenced by the dominant phenotype in parents. If Y represents the tall and y represent the short, then three genotypes are tall, and one is short. Expand above to include two traits in each parent:

Consider:

Y: dominant allele for tall
y: recessive allele for short
W: Dominant allele for white
w: recessive allele for black
The output as per Mendel would be combination of YyWw x YyWw, Table (1.2).

Table (1.2): Crossing between dominant and recessive alleles

	YW	Yw	yW	yw
YW	YWYW (tall)	YWYw (tall)	YWyW (tall)	YWyw (tall)
Yw	YwYW (tall)	YwYw (tall)	YwyW (tall)	Ywyw (tall)
yW	yWYW (tall)	yWYw (tall)	yWyW (short)	yWyw (short)
yw	ywYW (tall)	ywYw (tall)	ywyW (short)	Ywyw (short)

Note that there are twelve tall, and only four shorts.
For simplicity, alphabetical and numerical figures will be used.
Let us expand the alleles to include a third characteristic dwarf and brown. Table (1.3) would be a combination of ABC123, and still the dominant is A in the father and 1 in the mother.

Table (1.3): Expanding traits

	A1	A2	A3	B1	B2	B3	C1	C2	C3
A1	A1A1	A1A2	A1A3	A1B1	A1B2	A1B3	A1C1	A1C2	A1C3
A2	A2A1	A2A2	A2A3	A2B1	A2B2	A2B3	A2C1	A2C2	A2C3
A3	A3A1	A3A2	A3A3	A3B1	A3B2	A3B3	A3C1	A3C2	A3C3
B1	B1A1	B1A2	B1A3	B1B1	B1B2	B1B3	B1C1	B1C2	B1C3
B2	B2A1	B2A2	B2A3	B2B1	B2B2	B2B3	B2C1	B2C2	B2C3
B3	B3A1	B3A2	B3A3	B3B1	B3B2	B3B3	B3C1	B3C2	B3C3
C1	C1A1	C1A2	C1A3	C1B1	C1B2	C1B3	C1C1	C1C2	C1C3
C2	C2A1	C2A2	C2A3	C2B1	C2B2	C2B3	C2C1	C2C2	C2C3
C3	C3A1	C3A2	C3A3	C3B1	C3B2	C3B3	C3C1	C3C2	C3C3

The recessive of the genotype is16 out of 81 populations. If there are two dominants in the father, and two dominant in the mother, the result would be one recessive and 80 dominant, Table (1.4).

Table (1.4): Two dominants in each of the parent

	A1	A2	A3	B1	B2	B3	C1	C2	C3
A1	A1A1	A1A2	A1A3	A1B1	A1B2	A1B3	A1C1	A1C2	A1C3
A2	A2A1	A2A2	A2A3	A2B1	A2B2	A2B3	A2C1	A2C2	A2C3
A3	A3A1	A3A2	A3A3	A3B1	A3B2	A3B3	A3C1	A3C2	A3C3
B1	B1A1	B1A2	B1A3	B1B1	B1B2	B1B3	B1C1	B1C2	B1C3
B2	B2A1	B2A2	B2A3	B2B1	B2B2	B2B3	B2C1	B2C2	B2C3
B3	B3A1	B3A2	B3A3	B3B1	B3B2	B3B3	B3C1	B3C2	B3C3
C1	C1A1	C1A2	C1A3	C1B1	C1B2	C1B3	C1C1	C1C2	C1C3
C2	C2A1	C2A2	C2A3	C2B1	C2B2	C2B3	C2C1	C2C2	C2C3
C3	C3A1	C3A2	C3A3	C3B1	C3B2	C3B3	C3C1	C3C2	C3C3

Only one recessive is found in 81 populations.

The number of dominants follows the equation:

$$D = (n_f \cdot n_m)^2 - ((n_f - n_{fd}) \cdot (n_m - n_{md}))^2 \qquad (1)$$

Where D   is the number of dominants.
    n_f is the number of alleles in father,
    n_m is the number of alleles in mother,
    n_{fd} is the number of dominants in father,
    n_{md} is the number of dominants in mother.

Maximizing equation (1) by differentiating and equal to zero, it follows that the larger the number of dominant genotypes is the smaller the number of recessive ones will be.

If the mother has no dominant allele, then equation (1) will depend only on the crossing between the number of alleles in both parents and the dominant allele in the father.

Equation (1) is equivalent to the Hardy-Weinberg equilibrium which is $p^2 + 2pq + q^2 = 100\%$. Equating both equations, one can get:

$$(n_f \cdot n_m)^2 = 100\% \qquad (2)$$
$$D = p^2 + 2pq \qquad (3)$$
$$((n_f - n_{fd}) \cdot (n_m - n_{md}))^2 = q^2 \qquad (4)$$

And therefore:

$$(p^2 + 2pq)/q^2 = (n_f \cdot n_m)^2 - ((n_f - n_{fd}) \cdot (n_m - n_{md}))^2 / ((n_f - n_{fd}) \cdot (n_m - n_{md}))^2 \qquad (5)$$

If the number of dominant alleles in the father or mother equals the number of traits, the dominator will be equal to zero, and therefore the whole population will be dominants.

Now consider a man and a woman from the Amazon Forest married to each other. Their dominant phenotypes are intact and undamaged due to the constant environment and the natural food which does not contain chemicals or pollutants. The product of the genotype population will all be dominant. However, if their first offspring is married to a tall and white man from Greenland, their offspring would probably be a mixture between dominant and recessive. But how can this offspring from the Amazon get married to that offspring from Greenland without overcoming transportations, obstacles and other related barriers? The question aroused that if there is no change in the environment and no change in social behavior, would genotypes have recessive traits? In other words, is there any evolution as per Darwinian theory if there is no change in the natural world? This is a simple way to prove that natural selection is mathematically impossible of producing evolutionary change, because there is no gene drift when Amazonians married each other.

To prove that the Darwin theory can not be accepted, let us assume that all of the population is married to a father of pure recessive alleles, for the worst case scenario, as shown in Table (1.5).

Table (1.5): Equal recessive genotypes of a father of pure recessive alleles

			A1	A2	A3	B1	B2	B3	C1	C2	C3
B2		A1	A1A1	A1A2	A1A3	A1B1	A1B2	A1B3	A1C1	A1C2	A1C3
B3		A2	A2A1	A2A2	A2A3	A2B1	A2B2	A2B3	A2C1	A2C2	A2C3
B4		A3	A3A1	A3A2	A3A3	A3B1	A3B2	A3B3	A3C1	A3C2	A3C3
C2		B1	B1A1	B1A2	B1A3	B1B1	B1B2	B1B3	B1C1	B1C2	B1C3
C3	x	B2	B2A1	B2A2	B2A3	B2B1	B2B2	B2B3	B2C1	B2C2	B2C3
C4		B3	B3A1	B3A2	B3A3	B3B1	B3B2	B3B3	B3C1	B3C2	B3C3
D2		C1	C1A1	C1A2	C1A3	C1B1	C1B2	C1B3	C1C1	C1C2	C1C3
D3		C2	C2A1	C2A2	C2A3	C2B1	C2B2	C2B3	C2C1	C2C2	C2C3
D4		C3	C3A1	C3A2	C3A3	C3B1	C3B2	C3B3	C3C1	C3C2	C3C3

equals
to

A1A1R	A1A2R	A1A3R	A1B1R	A1B2R	A1B3R	A1C1R	A1C2R	A1C3R
A2A1R	A2A2R	A2A3R	A2B1R	A2B2R	A2B3R	A2C1R	A2C2R	A2C3R
A3A1R	A3A2R	A3A3R	A3B1R	A3B2R	A3B3R	A3C1R	A3C2R	A3C3R
B1A1R	B1A2R	B1A3R	B1B1R	B1B2R	B1B3R	B1C1R	B1C2R	B1C3R
B2A1R	B2A2R	B2A3R	B2B1R	B2B2R	B2B3R	B2C1R	B2C2R	B2C3R
B3A1R	B3A2R	B3A3R	B3B1R	B3B2R	B3B3R	B3C1R	B3C2R	B3C3R
C1A1R	C1A2R	C1A3R	C1B1R	C1B2R	C1B3R	C1C1R	C1C2R	C1C3R
C2A1R	C2A2R	C2A3R	C2B1R	C2B2R	C2B3R	C2C1R	C2C2R	C2C3R
C3A1R	C3A2R	C3A3R	C3B1R	C3B2R	C3B3R	C3C1R	C3C2R	C3C3R

One can see that, as per Mendel's laws, the number of dominant genotypes are still the same. In the table above, there are even more recessive components that have been added to the favor of Darwin theory. Also, if the same recessive components are added, the number of dominants will be increased, as shown in equation (5) if it is multiplied by q:

$$q.(p^2+2pq)/ q^2 = q . ((n_f . n_m)^2 - ((n_f-n_{fd}) . (n_m-n_{md})))^2 / ((n_f-n_{fd}) . (n_m-n_{md}))^2$$

$$= (p^2+2pq)/q$$

The denominator will be q instead of q^2, and therefore the percentage of dominants to recessives will be larger than before.

Inevitably, the calculations are suggestive of the genetic consistency within a given population if there were no intercultural marriages. Furthermore, environmental changes and cross-cultural engagements have only recently been realized compared to the length of time; 3 - 7 million years it takes for evolution to transform according to evolutionists, [http://www.drelser.com/shop/article_2/Heredity%2C-Darwin-and-misleading-theory.html?shop_param=cid%3D1%26aid%3D2%26]

1.7.2 Gene Regulation

Virtually every cell in our body contains a complete set of genes. During development different cells express different sets of genes in a precisely regulated fashion. Gene regulation occurs at the level of transcription of DNA to mRNA and at the level of translation of mRNA to protein. Genetic information (regulation) always goes from DNA to RNA to Protein.

The best study of gene regulation has been performed on E. coli bacteria that live in the colon of our body.

E. coli bacteria have a metabolic pathway that allows the synthesis of amino acid tryptophan (Trp). When tryptophan is not available in the media, then E. coli manufactures its own amino acids.

The pathway starts with a precursor molecule and proceeds through a cluster of five genes (enzymes) to manufacture the amino acid tryptophan. All five genes are transcribed together as a unit called an operon, which produces a single long piece of mRNA for all the genes.

Since the amount of Trp available from the environment varies considerably, E. coli controls the rate of Trp synthesis. So, with little Trp in a meal, E. coli compensates by making more, and vice versa.

There are two ways of regulating the amount of Trp:
- RNA polymerase binds to the operon for the synthesis of Trp
- RNA polymerase does not bind to the operon for inhibiting the synthesis of Trp.

The operon has three components, Figure (1.26).

- There are five genes which contain the genetic characteristics
- There is one promoter where the RNA polymerase binds to the DNA segment for the transcription.
- There is one operator which is the DNA segment found between the promoter and structural genes. It governs if the transcription will take place when the gene is ON or OFF.

Figure (1.26): Trp operon

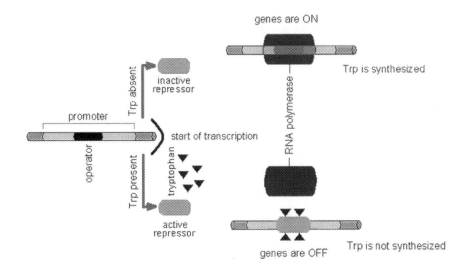

Figure (1.26) shows the two ways of gene regulation:

- The operon is ON where the RNA polymerase binds to the promoter, and the five genes are transcribed to one mRNA strand. The mRNA will then translate into the enzymes that control the Trp synthesis pathway.

- The pathway is OFF by a specific protein called the repressor. The repressor is inactive and can not bind to the operator. The Trp can not be synthesized.

The location and condition of the regulators, promoter, operator and structural DNA sequences is shown in Figure (1.27).These segments can determine the effects of common mutations.

Figure (1.27): Structural of DNA sequence

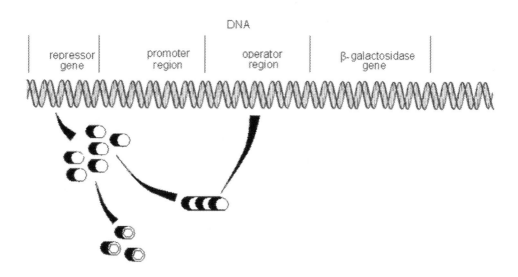

In effect, there are two pathways of the Trp: anabolic and catabolic.

- The anabolic is when theTrp is synthesized
- The catabolic is when the Trp is not synthesized, during which the sugar lactose is metabolized (catabolized).

For the metabolization of sugar lactose, E. coli makes an enzyme called beta galactosidase which is coded by the Lac Z gene, which is also produced by E. coli itself. The Lac Z gene is clustered on the same DNA in what is called the Lac Z operon. The Lac operon has three structural genes, the promoter (DNA segment where RNApolymerase binds and starts transcription), and the operator which turns ON and OFF similar to the Trp operon. So, if the sugar lactose is available in the media, the operator turns OFF.

It is concluded that the gene can be regulated between ON and OFF states. The Trp operon is an example of repressible gene regulation (anabolic) and the Lac operon is an example of inducible gene regulation (catabolic).

1.7.3 Genetic Crosses

Genetic recombination crossing over is the mutual exchange of the corresponding parts of the adjacent paternal and maternal chromatids of the pachytene of meiosis. In genetic crosses, crossover is a genetic operator used to vary the programming of a chromosome or chromosomes from one generation to the next. Genetic crosses are also used for producing new combinations of genes. The chromatids resulting from the interchange of segments are known as the cross over recombinants. The chromatids that remain intact are called non-crossover parental chromatids.

In vivo, many crossover techniques exist for organisms which use different data structures during the meiosis and mitosis cycles. There are several types of gene crosses:

1. Single Crossover

A single crossover point on both parents' organism strings is selected. All data beyond that point in either organism string is swapped between the corresponding parts of the adjacent paternal and maternal chromatids.The resulting offspring is shown in Figure (1.28).

Figure (1.28): Single point gene crossover

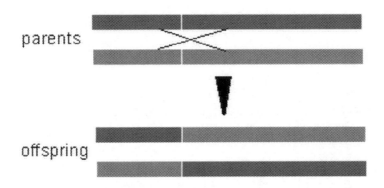

2. Double Point Crossover

A double point crossover is two points on the parent organism strings. Everything between the two points is swapped between the parent organisms, rendering two offspring, Figure (1.29).

Figure (1.29): Double point crossover

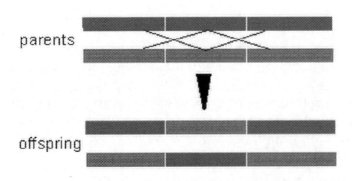

3. Cut and Splice Crossover

Cut and splice cross over results in a change in the length of the children strings. The reason for this difference is that each parent string has a separate choice of crossover point, Figure (1.30).

Figure (1.30): Cut and splice crossover

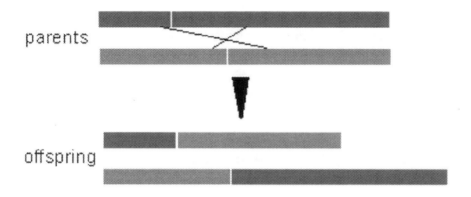

4. Chromatids Crossover

Crossing over is the exchange of strictly homologous segments between non sister chromatids of homologous chromosomes. Chromosomal crossover is an exchange of genetic material between homologous chromosomes. It is one of the final phases of genetic recombination, which occurs during prophase 1 of meiosis (diplotene) in a process called synapsis. Synapsis begins before the synaptonemal complex develops, and is not completed until near the end of prophase 1. Crossover usually occurs when matching regions on matching chromosomes break and then reconnect to the other chromosome. Crossing over occurs during the pachytene stage of meiosis and is responsible for recombination between linked genes. During the pachytene stage, each chromosome of a bivalent (chromosome pair) has two chromatids. Thus, each bivalent contains four chromatids or strands (four strand stages). Generally, one chromatid from each homologue is involved in crossing over. In the crossing over a segment of one chromatid becomes attached in place of the homologous segment of the non sister chromatid and vice versa.

Crossing over involves the breakage of two homologous chromosomes at precise points in the two non sister chromatids and then exchange of parts. This produces an X-like structure at the point of exchange of the chromatid segments. This structure is called a chiasma (plural chaismata). Chaismata occur more or less randomly. The mechanism of crossing over occurs randomly. Figure (1.31) shows the mechanism of chromatids crossing over.

Figure (1.31): Chromatids crossing over

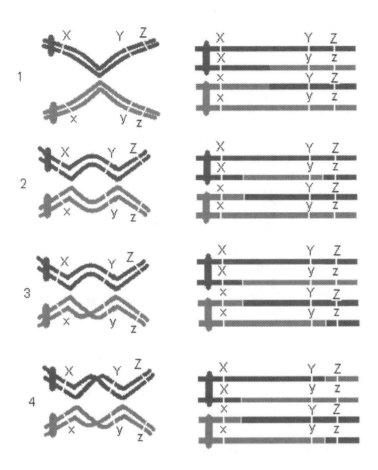

Chapter 2

Brain and Intelligence

2. The Brain

The brain is extremely complex. The cerebral cortex of the human brain contains roughly 15–33 billion neurons. The cerebrum is the largest part of the brain, accounting for 85 percent of the organ's weight. The distinctive, deeply wrinkled outer surface is the cerebral cortex, which consists of gray matter. Beneath this lies the white matter.

The neurons are linked with up to 10,000 synaptic connections each. Each cubic millimeter of cerebral cortex contains roughly one billion synapses. These neurons communicate with one another by means of long protoplasmic fibers called axons, which carry trains of signal pulses called action potentials (measured in mill voltages and micro voltages) to distant parts of the brain or body and target them to specific recipient cells.

It is very hard to imagine how all the complex intellectual capabilities conferred by the human brain, can develop from an embryo which begins from a single cell or rather the first two cells that create it.

The human nervous system starts to form very early in the embryo's development. At the end of the gastrulation phase, an elongated structure, the notochord, is laid down. The embryo thereby changes from a circular organization to an axial one.

In a mammalian embryo, the neural tube is initially a straight, linear structure as seen in the shape below:

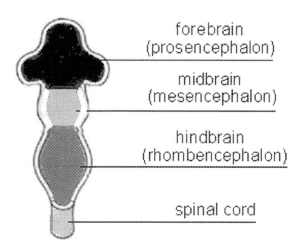

forebrain
(prosencephalon)

midbrain
(mesencephalon)

hindbrain
(rhombencephalon)

spinal cord

 At the start of the 4th week, this end of the neural tube begins to curve and divides into three bulges, known as the primitive vesicles. The forebrain, the midbrain, the hindbrain and the spinal cord start to emerge from the neural tube, see configuration below:

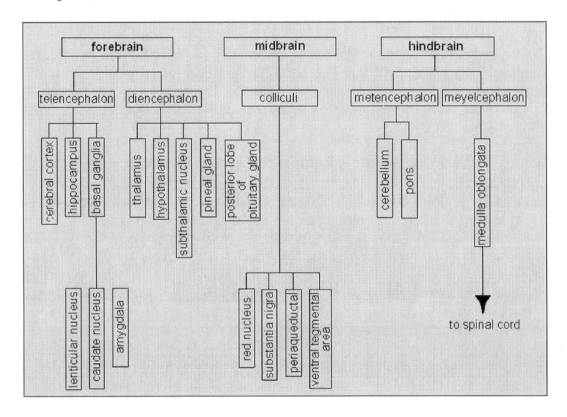

The human brain is a complex organ that controls and allows us to think, move, feel, see, hear, taste, and smell. It controls our body, receives and sends information, analyzes information, and stores information (memories).

The brain produces electrical signals which, together with chemical reactions, let the parts of the body communicate. Nerves send these signals throughout the body.

The brain is only 2% of the body's weight, but uses 20% of the oxygen supply and gets 20% of the blood flow. Blood vessels including arteries, capillaries, and veins, supply the brain with oxygen and nourishment, and take away wastes. If brain cells do not get oxygen for 3 to 5 minutes, they begin to die. The brain begins to suffer damage only after approximately 6 minutes of hypoxia (lack of oxygen).

2.1 Structure of the Brain

The brain has two cerebral hemispheres. Each takes care of one side of the body, but the controls are crossed: the right hemisphere takes care of the left side, and vice versa, Figure (2.1). Each hemisphere appears to be specialized for some behaviors. For example, it appears that the right side of the brain is dominant for spatial abilities, face recognition, visual imagery and music. The left side of the brain may be more dominant for calculations, math and logical abilities. Of course, these are generalizations. In normal people, the two hemispheres work together, are connected, and share information through the corpus callosum. Much of what we know about the right and left hemispheres comes from studies in people who have had the corpus callosum split in which surgical operation isolates most of the right hemisphere from the left hemisphere. This type of surgery is performed in patients suffering from epelipsy. The corpus callosum is cut to prevent the spread of the "epileptic seizure" from one hemisphere to the other.The hemispheres communicate with each other through a thick band of 200-250 million nerve fibers called the corpus callosum. (A smaller band of nerve fibers called the anterior commissure also connects parts of the cerebral hemispheres).

Figure (2.1): Cerebral hemispheres of the brain

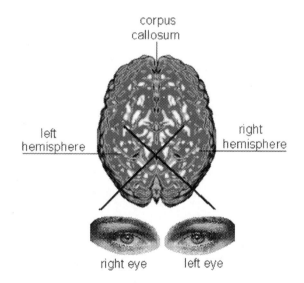

The brain is divided into three parts, as follows, Figure (2.2):

1) Forebrain: the telencephalon and diencephalon
2) Midbrain: the mesencephalon
3) Hindbrain: the metencephalon and myelencephalon

Figure (2.2): Parts of the brain

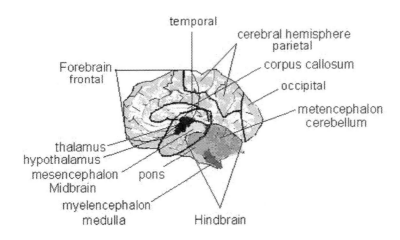

2.2 Function of the Brain

There are several different regions of the human brain, each with interacting but distinct functions. Table (2.1) shows the regions with their functions.

Table (2.1): Regions of the brain and their functions

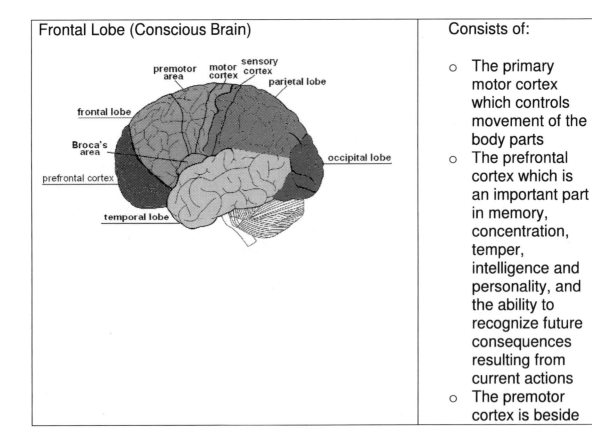

Frontal Lobe (Conscious Brain)	Consists of:
	o The primary motor cortex which controls movement of the body parts
	o The prefrontal cortex which is an important part in memory, concentration, temper, intelligence and personality, and the ability to recognize future consequences resulting from current actions
	o The premotor cortex is beside

	the primary motor cortex and guides our eye and head movements and sense of orientation ○ Broca's area is important in language production, see the figure below: The executive functions of the frontal lobes involve the ability to recognize future consequences resulting from current actions, to choose between good and bad actions, control voluntary movement, thinking, and feeling, override and suppress unacceptable social responses, and determine similarities and differences between things or events.
Parietal Lobe	The parietal lobe plays important roles in integrating sensory information from various parts of the body, knowledge of numbers and their relations, and in the manipulation of objects.
Occipital Lobe	○ Contains the visual cortex ○ The visual cortex contains association areas that help in the visual recognition of shapes and

	colors.
	The right side of the brain 'sees' the left visual space, whereas the left side of the brain 'sees' the right visual space The primary visual cortex manages vision and has a full map of the visual world.
Temporal Lobe	<ul><li>contains the auditory cortex helps that receives signals from the ear and lets us hear sounds and associate meanings with sounds</li><li>the Wernicke's area is important for language, speech and meaning</li></ul>The temporal lobe is involved in auditory perception and is home to the primary auditory cortex. It is also important for the processing of semantics in speech, hearing and vision. It contains the hippocampus and plays a key role in the formation of long-term memory.
Prefrontal Cortex	It is responsible for planning complex cognitive behaviors, personality expression,

	decision making and moderating correct social behavior. It helps focus attention, and gives meaning to perceptions.
Parietal Lobe	Parietal lobe coordinates signals received from other brain regions to interpret general sensory signals. It coordinates visual, auditory and language mechanisms , motor and sensory signals along with memory helps to identify objects
Thalamus	The major role of thalamus is to gate (relay) and otherwise modulate the flow of information (except smell) to cortex. For example, visual information from the retina is not sent directly to visual cortex but instead is relayed through the lateral geniculate nucleus of the thalamus.
Hypothalamus	One of the most important functions of the hypothalamus is to link the nervous system to the endocrine system via the pituitary gland. ITregulates basic biological drives, hormonal levels, sexual behavior, and controls

In the Thalamus row, a diagram appears with the following labels: pineal gland, caudate nucleus, thalamus, hippocampus, amygdala, pituitary gland, hypothalamus.

	autonomic functions such as hunger, thirst, and body temperature.
Pituitary Gland	It produces growth hormone, prolactin - to stimulate milk production after giving birth, ACTH (adrenocorticotropic hormone) - to stimulate the adrenal glands, TSH (thyroid-stimulating hormone) - to stimulate the thyroid gland, FSH (follicle-stimulating hormone) - to stimulate the ovaries and testes, and LH (luteinizing hormone) - to stimulate the ovaries and testes.
Hippocampus	It helps regulate emotion and memory. Functionally, the hippocampus is part of the olfactory cortex, that part of the cerebral cortex essential to the sense of smell. It mediates learning and memory formation.
Amygdala	It is a complex structure involved in a wide range of normal behavioral functions and psychiatric conditions. It is responsible for anxiety, emotion, and fear.

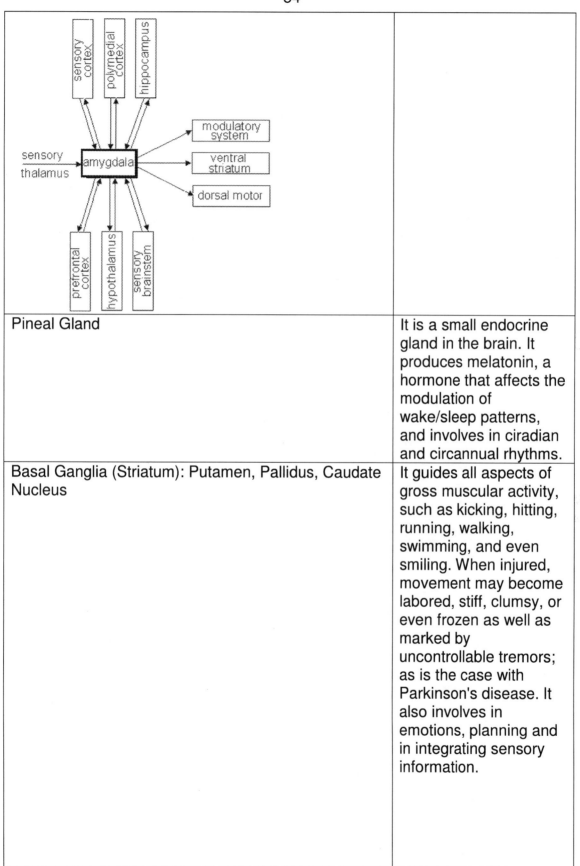

Pineal Gland	It is a small endocrine gland in the brain. It produces melatonin, a hormone that affects the modulation of wake/sleep patterns, and involves in ciradian and circannual rhythms.
Basal Ganglia (Striatum): Putamen, Pallidus, Caudate Nucleus	It guides all aspects of gross muscular activity, such as kicking, hitting, running, walking, swimming, and even smiling. When injured, movement may become labored, stiff, clumsy, or even frozen as well as marked by uncontrollable tremors; as is the case with Parkinson's disease. It also involves in emotions, planning and in integrating sensory information.

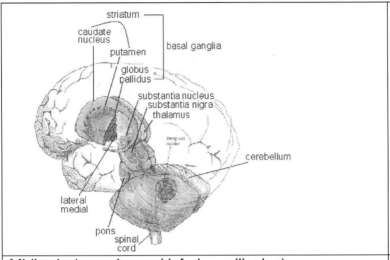

| Midbrain (superior and inferior colliculus) | The midbrain (mesencephalon) is considered part of the brain stem. Its substantia nigra is closely associated with motor system pathways of the basal ganglia. It relays sensory information from the spinal cord to the forebrain. |

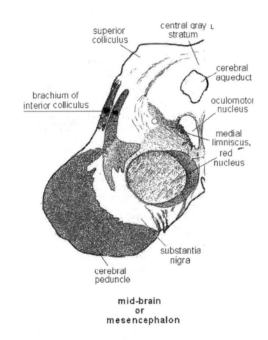

| Medulla | The medulla contains the cardiac, respiratory, vomiting and vasomotor centers and deals with autonomic functions, such as breathing, heart rate and blood pressure. |

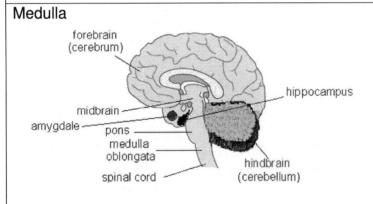

midbrain medulla eye lacrimal gland mucous memb., nose and palate submaxillar gland sublingual gland mucous membrane, mouth parotid gland heart larynx, trachea, bronchi esophagus stomach blood vessels liver pancreas small intestine large intestine	
Pons	The pones is located above the medulla. It contains nuclei that relay signals from the cerebrum to the cerebellum, along with nuclei that deal primarily with sleep, respiration, swallowing, bladder control, hearing, equilibrium, taste, eye movement, facial expressions, facial sensation, and posture. The nucleus of the pones regulates the change from inspiration to expiration.
Cerebellum	The cerebellum is involved in the coordination of voluntary motor movement, balance and equilibrium and muscle tone. The cerebellum does not initiate movement, but it contributes to coordination, precision, and accurate timing. It

	receives input from sensory systems and from other parts of the brain and spinal cord, and integrates these inputs to fine tune motor activity.
Limbic System	The limbic system is a complex set of structures that lies on both sides of the thalamus, just under the cerebrum. It includes the hypothalamus, the hippocampus, the amygdala, and several others (cingulate gyrus, ventral tegmental area, basal ganglia, prefrontal cortex) nearby areas. It appears to be primarily responsible for our emotional life, and has a lot to do with the formation of memories.
Cingulate Gyrus	It is an integral part of the limbic system, which is involved with emotion formation and processing, learning, and memory, and is also important for executive function and respiratory control.
Reticular Formation	The reticular formation is a comprehensive network of nerves that is found in the central area of the brainstem. The

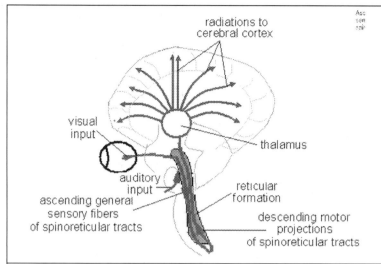

radiations to cerebral cortex

visual input

thalamus

auditory input

reticular formation

ascending general sensory fibers of spinoreticular tracts

descending motor projections of spinoreticular tracts

functions of the reticular formation involve many of the essential functions of the body, such as the ability to obtain recuperative sleep, sexual arousal, and many other things such as somatic motor control, cardiovascular control, Pain modulation, sleep and consciousness, and habituation.

Ventricles and Central Canal

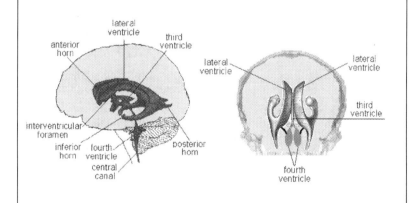

lateral ventricle

anterior horn

third ventricle

lateral ventricle

lateral ventricle

interventricular foramen

inferior horn

fourth ventricle

central canal

posterior horn

third ventricle

fourth ventricle

The functions of the ventricles of the brain is the protection of the brain by providing cushioning. The cerebrospinal fluid (CSF) produced in the ventricles act as the cushion to minimize the impact of any kind of trauma to the head. The CSF travels through the ventricles system provides a pathway and provides protection to the brain. CSF is concerned with the excretion of waste products such as, harmful metabolites or drugs from the brain, besides transporting the hormones to various part of the brain. It also provides buoyancy to the brain, which in turn, helps to reduce the weight of the brain. Just because our brain remains immersed in cerebrospinal fluid, its weight reduces from

	1,400 gm to almost 50 gm, which in turn, reduces pressure at the base of the brain.
Corpus Callosum 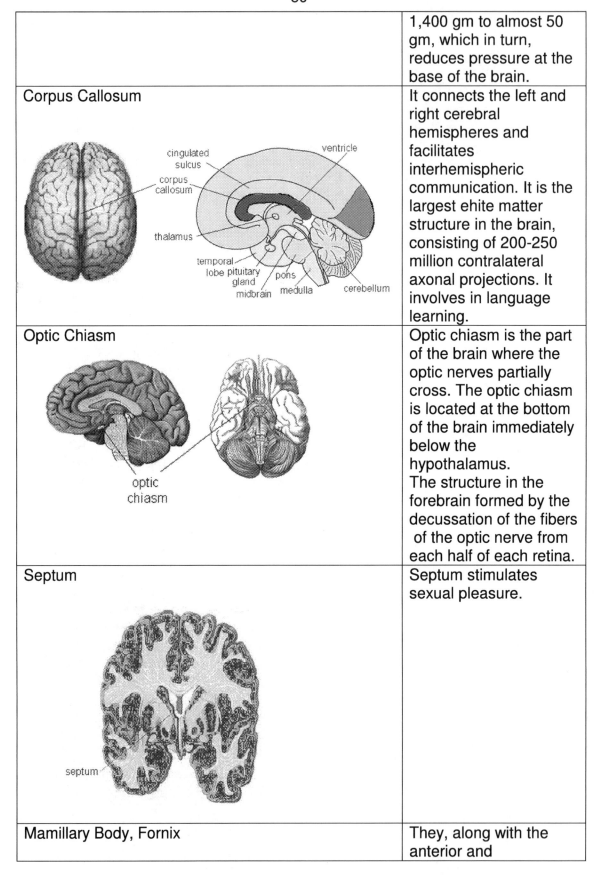	It connects the left and right cerebral hemispheres and facilitates interhemispheric communication. It is the largest ehite matter structure in the brain, consisting of 200-250 million contralateral axonal projections. It involves in language learning.
Optic Chiasm	Optic chiasm is the part of the brain where the optic nerves partially cross. The optic chiasm is located at the bottom of the brain immediately below the hypothalamus. The structure in the forebrain formed by the decussation of the fibers of the optic nerve from each half of each retina.
Septum	Septum stimulates sexual pleasure.
Mamillary Body, Fornix	They, along with the anterior and

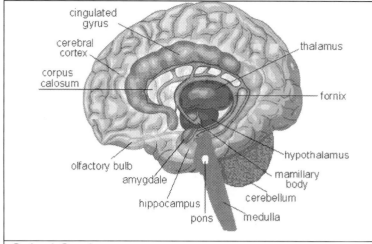

cingulated gyrus
cerebral cortex
corpus calosum
thalamus
fornix
hypothalamus
olfactory bulb
amygdale
mamillary body
cerebellum
hippocampus
medulla
pons

dorsomedial nuclei in the thalamus, are involved with the processing of recognition memory. They are believed to add the element of smell to memories. They have a role in emotional behavior, learning, and motivation.

Spinal Cord	<ul><li>The spinal cord receives information from skin, joints, and muscles</li><li>Sends back signals for both voluntary and reflex movements</li><li>Ttransmits signals from internal organs to the brain and from the brain to internal organs</li><li>Connects the brain to peripheral organs and tissue</li><li>Iin addition, the spinal cord contains</li></ul><ul><li>Ascending pathways through which sensory information reaches the brain</li><li>Descending pathways that</li></ul>

	relay motor commands from the brain to motor neurons
Functions of Brain's Parts 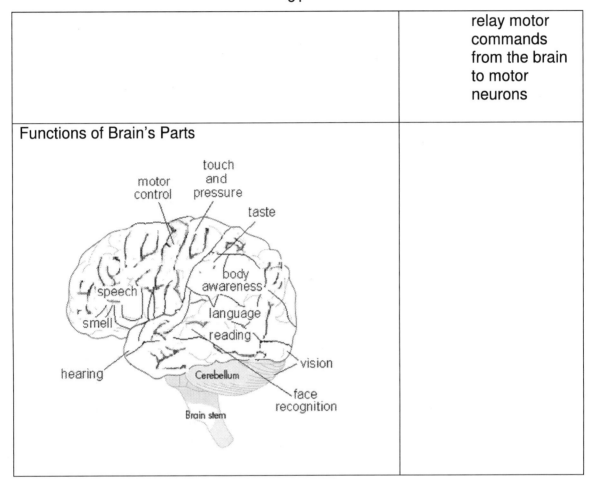	

2.3 Neurons and Nerves

The human nervous system consists of two main systems: the central nervous system (CNS), and the peripheral nervous system (PNS), which includes the somatic motor nervous system, and the sensory nervous system. The central nervous system contains the majority of the nervous system and consists of the brain and the spinal cord, as well as the retina. The main function of the PNS is to connect the CNS to the limbs and organs. The peripheral nervous system is divided into the somatic nervous system and the autonomic nervous system.

2.3.1 Neuron

The neuron consists of two portions: the cell body and the axon. The cell body is like the other cells. It contains a nucleus and cytoplasm. It is different from other cells because out of the cell body, long threadlike projections protrude. These are called dendrites ("tree" in Greek). At one point of the cell, however, there is a particularly long extension that usually does not branch throughout most of its sometimes enormous length. This is the axon (the axis), Figure (2.3).

Figure (2.3): Neuron

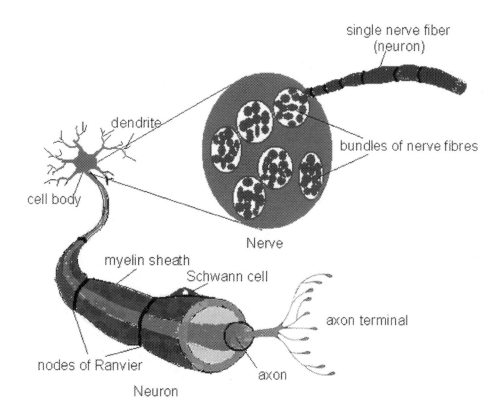

single nerve fiber
(neuron)

dendrite

bundles of nerve fibres

cell body

Nerve

myelin sheath

Schwann cell

axon terminal

nodes of Ranvier

axon

Neuron

Structurally, a neuron consists of dendrites, axon and axon terminal. Dendrites conduct nerve impulses toward the cell body. The axon conducts nerve impulses away from the cell body through the axon terminal. To speed up the transmission and to keep the signal from scattering and propagation (like an electrical cable), the axon is sheathed with a myelin layer that is made up of Schwann cells. Messages move as fast as 400 km per hour.

The axon terminal receives messages from the cell body, and then transmits the messages to neighboring neurons via the release of neurotransmitters which are endogenous chemicals that relay, amplify, and modulate signals between neurons and other cells, Figure (2.4).

Figure (2.4): Sending and receiving messages through neurons

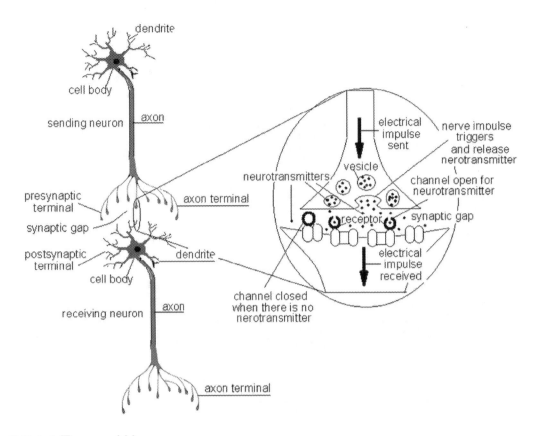

2.3.1.1 Types of Neurons

The three basic types of neurons are the motor neuron (efferent), the sensory neuron (afferent), and the interneuron. The motor neurons are specified to send messages away from the Central Nervous System. The sensory neurons are specified in the senses of taste, touch, hearing, smell, and sight. They send messages from the sensory receptors to the Central Nervous System. The inter neurons are sort of a mix of both a sensory neuron and a motor neuron, Figure (2.5).

64

Figure (2.5): Sending messages to and from the Central Nervous System

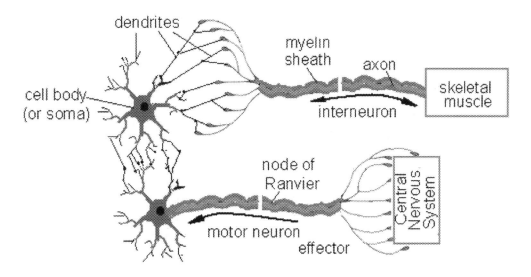

2.3.1.2 Transmitting Messages into the Neuron

At rest, there is an electrical charge difference between the inside and outside of the neuron because of either positively or negatively charged ions that are caused by sodium (Na^+), potassium (K^+) and chloride (Cl^-). The inside of the neuron is more negatively charged than the outside of the neuron (because sodium is more than ten times more concentrated outside the neuron's membrane than inside of the neuron), and the neuron is said to be polarized, i.e., there is a difference in electrical charge between the inside and outside of the neuron.

The neuron has channels that can permit chemicals to pass into and out of the neuron. The sodium channels are completely closed during the resting potential, but the potassium channels are pertly open, so potassium can flow slowly out of the neuron.

The protein of the neuron loves the potassium and hates the sodium, thus, the neuron pumps out the sodium and pumps in the potassium. Because the sodium atom ends with one electron and the chlorine in the sodium chloride (salt) ends with seven electrons, the sodium atom looses the electron to the chlorine atom. Thus the sodium atom becomes a positive ion (cation), and the chlorine atom becomes negative ion (anion), Figure (2.6).

Figure (2.6): Neurons and dynamical polarization

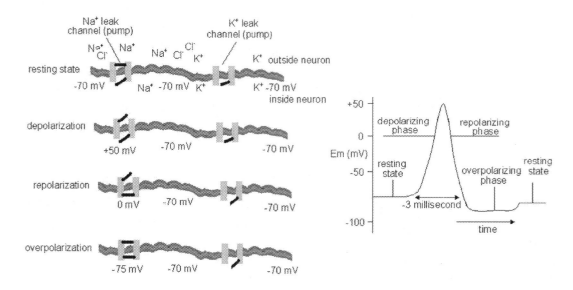

Substances that act as neurotransmitters can be broadly classified into three major groups:

1. Amino acids such as glumatic acid (glutamate), GABA, aspartic acid and glycine
2. Peptides such as vasopression, somatostatin, and neurotension
3. Monoamines such as neropinophrene, dopamine serotonin and acetylcholine

The central machine of the brain's neurotransmitters is glutamate and GABA. Some examples of neurotransmitter action:

- Dopamine – voluntary movement
- Serotonin – sleep and temperature regulation
- GABA (gamma aminobutryic acid) – motor behaviour
- Glycine – spinal reflexes and motor behavoiur
- Noradrenaline – wakeful and arousal
- Acetycholine – voluntary movement of the muscles
- Neuromodulator – sensory transmission (pain)
- Enkephalin (opiate) – stress, pain killer, promote calm
- ATP – energy
- Insulin - sugar

Figure (2.7) shows the effects on the mental states induced by three major neurotransmitters.

Figure (2.7): Effects of dopamine, serotonin and noradrenaline onto the mental states of the brain

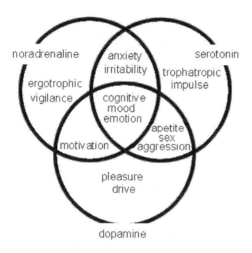

2.4 Brain Intelligence

The basic mechanisms by which the brain produces complex phenomena, like consciousness and intelligence are still poorly understood. Some of the anatomical variables that have been studied (using MRI) in association with psychometric test scores include total brain volume, the size and shape of the frontal lobes, the amount of grey and white matter, and the overall thickness of the cortex.

2.4.1 Brain Size

People with bigger brains are smarter than their smaller-brained counterparts, according to a study conducted by a Virginia Commonwealth University researcher published in the journal "Intelligence." The study is the most comprehensive of its kind, drawing conclusions from 26 previous – mostly recent – international studies involving brain volume and intelligence. It was only five years ago, with the increased use of MRI-based brain assessments, that more data relating to brain volume and intelligence became available. The brain consumes about 25% of the body's metabolic energy in some species. Because of this fact, larger brains are associated with higher intelligence. Brain size in vertebrates may relate to social rather than mechanical skill. Cortical size relates directly to a pairbonding life style. Among primates cerebral cortex size varies directly with the demands of living in a large complex social network, [Dunbar RI, Shultz S (2007-09-07). "Evolution in the social brain". *"Science"* 317: 1344–1347. doi:10.1126/science.1145463. PMID17823343].

A study on twins (Thompson *et al.*, 2001) showed that frontal gray matters volume was correlated with the General Intelligence Factor (psychologists use the term "g", for general intelligence factor) and highly heritable. When people are examined on a variety of tests, spanning different cognitive abilities – verbal ability, spatial reasoning, abstract logic, memory – it is found that people who do well on one of these tests tend to also do well on the others. A related study has reported that the correlation between brain size (reported to have a heritability of 0.85) and g is 0.4, and that correlation is mediated entirely by genetic factors (Posthuma et al. 2002). Results from a large number of twins, family and adoption studies agree that the heritability of g is very high – at least 50% and perhaps as high as 80%. (This means that 50-80% of the variance in g across the population is due to differences in genes).

Effects of a shared family environment are seen at early ages but these tend to disappear when examined in older individuals. Whatever the effect of the family environment on IQ measures when an individual is within that environment, these effects seem to be temporary and diminish in later life.

While a correlation with overall brain size has been repeatedly noted, this leaves a lot of the variance in intelligence unexplained (and is also not particularly informative). Is intelligence localized to a certain brain region or is it a distributed property of the entire network?

Comparing the ratio of brain weight to body weight of different species, the human has the largest ratio (2.1), except for the mouse which is 3.2. All species have a ratio of less than 1.

Our memory is a complex phenomenon, however, and of course it involves other parts of the brain as well. It is a mixture of psychology and neuroscience. In psychology, memory is an organism's ability to store, retain, and recall information. In recent decades, it has become one of the principal pillars of a branch of science called cognitive neuroscience, an interdisciplinary link between cognitive psychology and neuroscience.

There are three types of memory:

- Sensory memory
- Short-term memory
- Long-term memory

In memory, it also seems increasingly likely that these various types of memory bring different parts of the brain into play.

Sensory memory takes the information provided by the senses and retains it accurately but very briefly. Sensory memory corresponds approximately to the initial 200 - 500 milliseconds after an item is perceived. Nevertheless, it represents an essential step for storing information in short-term memory which then leads to long-term memory as per the following flow chart:

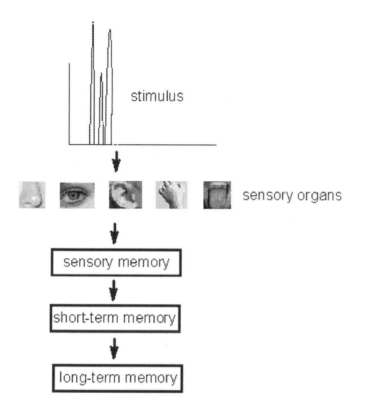

2.5 Neuroanatomy

The human nervous system is divided into the central and peripheral nervous systems.

The central nervous system (CNS) consists of:

- The brain
  1. Rhombencephalon or hindbrain which consists of medulla, pons and cerebellum
  2. Mesencephalon or midbrain
  3. Prosencephalon or forebrain which consists of the diencephalon and the telencephalon
  4. Cerebral hemisphere which consists of many smaller parts

- The spinal cord
  The peripheral nervous system (PNS) is composed of nerves and ganglia. A ganglion is a collection of neuronal cell bodies outside the CNS. Typically a ganglion is a lump on a nerve, but many of the ganglia associated with internal organs are of microscopic size. The peripheral nervous system is made up of all neurons in the body outside of the central nervous system, and includes:

1. The sensory somatic nervous system is made up of afferent neurons that convey sensory information from the sense organs to the brain and spinal cord, and efferent neurons that carry motor instructions to the muscles. The sensory-somatic system consists of 12 pairs of cranial nerves and 31 pairs of spinal nerves.

2. The autonomic nervous system consists of sensory neurons and motor neurons that run between the central nervous system (especially the hypothalamus and medulla oblongata) and various internal organs such as the:
   - Heart
   - Lungs
   - Viscera
   - Glands (both exocrine and endocrine)

The autonomic system has two groups:

   a- The sympathetic nervous system activates what has been called the "fight-or-flight" response that prepares the body for action
   b- The parasympathetic nervous system conserves energy and prepares the body to rest.
   The diagram of the human nervous system is shown in Figure (2.8).

Figure (2.8): Human nervous system

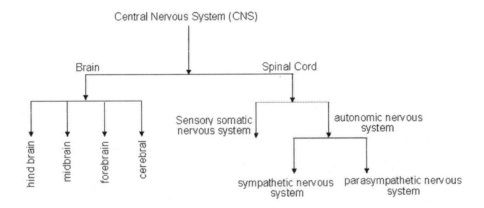

All of the pieces of information decoded in the various sensory areas of the cortex converge in the hippocampus, which then sends them back where they came from.

2.6 Memory

As we mentioned before there are three types of memory (sensory, short-term and long-term). The sequence of storing and retrieving the memory is like the following:

2.6.1 Short-Term Memory

First, the brain's cerebral cortex receives nerve messages from sensory organs such as eyes, ears, and touch sensors. This sensory stimulus is held for a fraction of a second in the sensory memory. Unless the person pays attention to the source of the message for about eight uninterrupted seconds to encode the stimulus into short-term memory, it will be lost, (Newton's Apple: memory). The memory then is stored on something similar to an electronic tape loop (although some scientists debate the existence of that loop). Once a complete loop is made, three things can happen: (1) the information can be "rehearsed" (repeated) silently or aloud, which will provide auditory cues; 2) the information goes into long-term memory; or 3) the information will be lost, Figure (2.9).

Figure (2.9): Storing and retrieving of memory

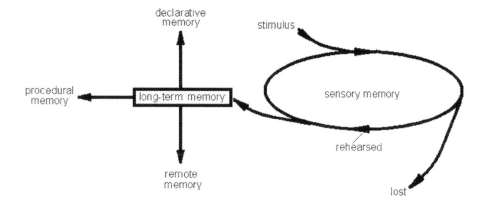

2.6.2 Long-term Memory

Long-term memory is that part of our memory storage system that has unlimited capacity to retain information over an extended time, (Newton's Apple: memory). At least three different types of memory are included in the long term memory. "Procedural memory represents motor or skill learning which is memory without verbal mediation and thus without record. It includes learning how to drive a car use key board of the computer, Figure (9). Such memories are slow to acquire but more resistant to change or loss. "Declarative memory is memory for facts, such as names and dates. It is fast changing, quick to acquire but quick to be lost".

"Remote memory simply refers to memories that were acquired early...They represent the foundation memories upon which more recent memories are built...Since early acquired information is the foundation for new memories and may be linked to many more new memories, such memory is less subject to change and/or loss", "memory", (from University of Memphis Neuropsychology Program).

Another representation of short-term and long-term memory depends on the acceptance of the hippocampus to the impulse. If the impulse is accepted, it will be stored in the long-term shelf; otherwise it will be lost or go to the short-term shelf, Figure (2.10).

Figure (2.10): Hippocampus' control of long and short-term memory

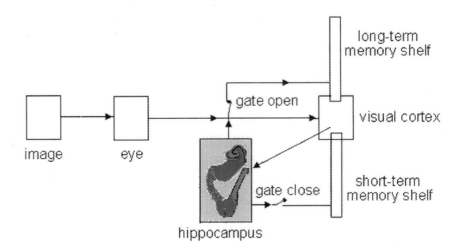

2.7 Changes in the Thickness of the Brain's Cortex

Shaw et al.'s (2006) report was focused on the structure of the brain's cortex (prefrontal cortex) and it development. This cortex is a thin layer located on the surface of the brain where most of the sophisticated information processing occurs. Before the age of seven years, the cortex continues to get thicker because of the growing neurons and its dendrites. After a certain age (varies in every child) a trimming process to the underused neurons and dendrites begins to occur, and the cortex starts to get thinner and the brain becomes more intelligent in the teen years. Researchers and psychologists discovered that the thickening and thinning process varied among children with different level of intelligence. The thinner the cortex is the more intelligent the person is. This was proved through scanning processes using functional magnetic resonance imaging (FMRI) and other sophisticated imaging equipment. Variations were particularly noticeable in the prefrontal cortex, located just behind the forehead. Figure (2.11) shows the changes in the thickness of the brain's cortex between ages 7 and 19. This prefrontal cortex is responsible for reasoning, planning and decision making.

Figure (2.11): Changes in the thickness of the brain's cortex

right lateral   right lateral

right medial   right medial

2.8 The Five Stages of the Human Brain (The Kübler-Ross model)

There are definite stages that the brain goes through when a mature person is coping with difficult and challenging experiences in his/her life. This may also include significant life events such as the death of a loved one, divorce, drug addiction, the onset of a disease or chronic illness, an infertility diagnosis, as well many tragedies and disasters. Through these stages (called Kübler-Ross model), the brain enables us to adjust to situations that might otherwise be too traumatizing. The stages are denial, anger, bargaining, depression, and finally the stage of acceptance.

2.8.1 Denial Stage

Denial can be very difficult for both family members and medical staff. For example, the head-injured person may say, "There's nothing wrong with me" or "Yes, I can drive," but family members who have been with the person know that it would be dangerous.

Five key issues are at stake in the human nature debate: the fear of inequality, the fear of imperfection, the fear of determinism, the fear of nihilism and the fear of human nature.

The fear of inequality arises from the idea that if we're blank slates, we must be equal. That follows from the mathematical truism that zero equals zero equals zero. But if the mind has any innate organization, according to this fear, then different races, sexes, or individuals could be biologically different, and that would condone discrimination and oppression.

The second fear is the fear of imperfection—the dashing of the ancient dream of the perfection of humankind. It runs more or less as follows: if ignoble traits such as selfishness, violence, or prejudice are innate, that would make them unchangeable, so attempts at social reform and human improvement would be a waste of time. Such bad traits should be tolerated.

2.8.2 Anger Stage

Denial is a very common problem, but eventually it breaks down to the next stage which is anger and depression. Anger is an emotion. The physical effects of anger include increased heart rate, blood pressure, and levels of adrenaline and noradrenaline. The external expression of anger can be found in facial expressions, body language, physiological responses, and at times in public acts of aggression. Sometimes anger is a way to shield ourselves from feeling intense pain; other times it's the simple contrast between other peoples' concerns and the sheer magnitude of what we're going through that triggers an attack of bitterness or frustration. Anger can cause sudden attacks of self-pity and frustration or bursts of outrage and a sense of injustice, and bitterness or resentment.

2.8.3 Bargaining Stage

After the initial two stages of denial and anger, the brain gets into the bargaining stage, where concepts like "What ifs" and "If Only" hypotheses crowds the thinking generated from a desperate attempt to regain control. The depression stage creates an obsessive and even a destructive stage of grief for the brain, especially if the person "gets stuck" in this stage, resulting in overwhelming anxiety and obsessive-compulsive symptoms. In the bargaining stage, people may ask or "bargain" for the impossible (that they will be the same as they were) as a way to deal with the true reality.

2.8.4 Depression Stage

People in the depression stage will become so overwhelmed with their situation that they will give up on trying to do anything. After the bargaining stage, the brain goes into the depression mood. They don't see any solution to the problem. Instead they turn in towards themselves, they turn away from any solution and any help that others can give them. Depression may be seen in a number of passive behaviors. In the workplace, in the home, and within the family and relatives, depression can also appear in tearful and morose episodes.

2.8.5 Acceptance Stage

Acceptance is typically visible by people taking ownership both for themselves and their actions. The individual begins to come to terms with their difficult and challenging experiences. They start to do things and take note of the results, and then change their actions in response. They will appear increasingly happier and more content as they find their way forward.

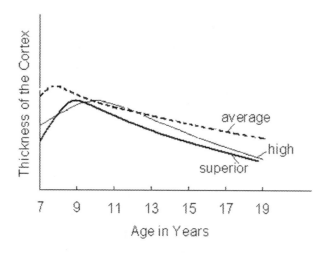

2.9 Neurobiology of Intelligence

There is strong evidence that the lateral prefrontal cortex and possibly other brain areas support intelligent behavior. Variations in intelligence and brain structure are heritable, but are also influenced by factors such as education, family environment, and environmental hazards.

Genetic research has demonstrated that:

- Intelligence levels can be inherited, and since genes work through biology, there must be a biological basis for intelligence.
- The Parieto-Frontal Integration Theory (P-FIT) identifies a brain network related to intelligence, one that primarily involves areas in the frontal and the parietal lobes (Richard Haier of the University of California, Irvine and Rex Jung of the University of New Mexico).
- The parietal lobes are responsible for spatial reasoning, visual processing and logic, and the corpus callosum pulls together information from both sides of the body.
- The quality of the myelin in the brain's white matter indicates the level of intelligence. If the water diffuses rapidly in a specific direction, it means that the brain has very fast connections. If it diffuses more broadly, that's an indication of slower signaling, and lowers intelligence (the brain has white and grey matter. The white matter is situated between the brainstem and cerebellum. The white matter consists of structures at the core of the brain such as the thalamus and hypothalamus. It relays sensory information from the rest of the body to the cerebral cortex, as well as in the regulation of autonomic unconscious functions, such as body temperature, heart rate and blood pressure. The grey matter includes

regions of the brain involved in muscle control, sensory perceptions, such as seeing and hearing, memory, emotions and speech. If the nervous system were a computer network, gray matter – a non-myelinated portion that contains nerve cells and capillaries – would be the computers and white matter the cables or the axons).

- From 30 previous – mostly recent – international studies involving brain volume and intelligence, it is concluded that the brain mass/volume is proportional to the intelligence. Some studies reveal that the ratio of the brain' mass to the body's mass is proportional to the intelligence.
- The brain and body are inseparably interwoven. The brain influences and is influenced by the farthest reaches of the body via the central and autonomic nervous systems. For example, the enteric nervous system in the abdomen, popularly known as the "second brain" in the gut, contains more neurons than the spinal cord. It sends messages to the brain far more often than it receives them, and can function without intervention by the central nervous system including the brain which is inside your skull (Gershon, 1999). Other nerves beyond innervation, neurotransmitters and neuropeptides have now been found in the immune system, heart, gut and connective tissue. (Pert 1997, Gershon 1994, Pearsall 1998) This establishes a direct neurochemical link between the brain and the rest of the body, beyond the previously known physical and mechanical neural pathways. Dr. Candace Pert, former researcher at the National Institute of Mental Health and the scientist who discovered the endorphin receptor (endorphins - pain killers - are released from the brain when severe injury occurs), refers to these neuropeptides as "bits of brain" that float throughout the body. This and related discoveries have Pert saying "I can no longer make a strong distinction between the brain and the body."

2.9.1 Brain Gene and Role in Intelligence

A multi-institution team led by University of Utah Professor Julie Korenberg says its findings could have implications for the understanding of intelligence and treatment of neurological disease in the general population. They've found the brain gene STX1A plays a significant role in the level of intelligence displayed by Williams Syndrome patients. They found variations in the expression of STX1A could account for 15.6 percent of cognitive variation in a group of 65 Williams Syndrome patients. STX1A is involved in the electrochemical processes that occur at the brain's synapses. Williams's syndrome is a rare neurodevelopment disorder caused by a deletion of about 26 genes from the long arm of chromosome 7… a tiny fraction of the nearly 30,000 genes found in humans. But such patients have one less copy each of the genes in question than the general population and typically exhibit an IQ of 60, compared to an average of 100 for the general population.

2.9.1.1 White and Gray Matters

Although there are essentially no disparities in general intelligence between the sexes, a UC Irvine study has found significant differences in brain areas where males and females manifest their intelligence. The study shows women having more white matter and men more gray matter related to intellectual skill, revealing that no single neuroanatomical structure determines general intelligence and that different types of brain designs are capable of producing equivalent intellectual performance. In general, men have approximately 6.5 times the amount of gray matter related to general intelligence than women, and women have nearly 10 times the amount of white matter related to intelligence than men. Gray matter represents information processing centers in the brain, and white matter represents the networking of – or connections between – these processing centers. If you compare the gray and white matters to a computer, the gray matters perform arithmetic and logic operations, whereas the white matters do the transmission and receiving (internal buses) between the gray matters. In the central nervous system, the "grey matter" is composed of the neurons' cell bodies and their dense network of dendrites. The grey matter includes the centre of the spinal cord and the thin outer layer of the cerebral hemispheres, commonly known as the cortex. The white matter consists of the myelin sheathing that covers the axons of these same neurons to enable them to conduct nerve impulses more rapidly. These myelinated axons are grouped into bundles (the equivalent of nerves) that make connections with other groups of neurons.

Gray matters correlate positively with various abilities and skills. White matters correlate with the speed with which we can process information.

2.9.1.2 Fluid and Crystallized Intelligence

Fluid intelligence (abbreviated Gf) is defined by Raymond Cattell as "...the ability to perceive relationships independent of previous specific practice or instruction concerning those relationships." Fluid intelligence is the ability to think and reason abstractly and solve problems inductively and deductively. This ability is considered independent of learning, experience, and education. It is the speed at which we can analyze information, the capacity of the working memory, and the ability to detect relationships among stimuli. Examples of the use of fluid intelligence include solving puzzles and coming up with problem solving strategies. Gf has been linked to a wide and diverse array of elementary cognitive processes during performance and includes:

Fluid Intelligence is a broad (stratum level II) ability in the Cattell-Horn-Carroll (CHC) theory of cognitive abilities. Gf refers to deliberate and controlled mental operations employed to solve novel (on the spot) problems that cannot be solved or performed automatically. In general, Gf mental operations may invoke drawing inferences, concept formation, classification, generating and testing hypothesis, identifying relations, comprehending implications, problem solving, extrapolating, and transforming information. Gf includes:

- General Sequential (deductive) Reasoning (RG), which is the ability to engage in a solution of a problem due to deductive reasoning and conclusions from given general conditions or premises to the specifics. This is also known as hypothetico-deductive reasoning.
- Induction (I), which is the ability to underline and combine separate pieces of information in the deduction of inferences, rules, hypothesis or conclusions.
- Quantitative Reasoning (RQ), which is the ability to inductively (I) and/or deductively (RG) reason with concepts involving mathematical relations and properties.
- Speed of Reasoning (RE), which is the ability to perform reasoning tasks (e.g., quickness in generating as many possible rules, solutions, etc., to a problem) in a limited time. Speed or fluency is the main factor of the performance.
- Piagetian Reasoning (RP), which is the ability to demonstrate the acquisition and application (in the form of logical thinking) of cognitive concepts as defined by Piaget's developmental cognitive theory. These concepts include seriating (organizing material into an orderly series that facilitates understanding of relationships between events), conservation (awareness that physical quantitatives do not change in amount when altered in appearance), classification (ability to organize materials that possess similar characteristics into categories), etc. The relation between RP with reasoning abilities measured by more conventional tests (I, RG, RQ) is not clear.

Crystallized intelligence (abbreviated Gc) is learning from past experiences and learning. It relies on accessing information from long-term memory. *Gc* is typically described as a person's wealth of acquired knowledge of the language, information and concepts of specific a culture, and/or the application of this knowledge. Situations that require crystallized intelligence include reading comprehension and vocabulary exams. This type of intelligence is based upon facts and rooted in experiences, good judgment, and mastery of social conventions. This type of intelligence becomes stronger as we age and accumulate new information. For example, the Wechsler Adult Intelligence Scale (WAIS) measures fluid intelligence on the performance scale and crystallized intelligence on the verbal scale. The overall IQ score is based on a combination of these two scales. Crystallized intelligence includes:

- Language Development, which is the ability to develop, understand and apply words, sentences, and paragraphs in spoken native language skills to express or communicate a thought or feeling.

- Lexical Knowledge, which is the ability to express vocabularies (nouns, verbs, or adjectives) that can be understood in terms of correct word (semantic) meanings. Although evidence indicates that vocabulary knowledge is a separable component from lexical knowledje, it is often difficult to disentangle these two highly corrected abilities in research studies.
- Verbal Information Range of general stored knowledge.
- Communication Ability which is the ability to speak in real life situations using oral production and fluency (e.g., lecture, conference, and group participation) in a manner that transmits ideas, thoughts, or feelings to one or more individuals.
- Foreign Language Proficiency and Foreign Language Aptitude which is the ability to learn and develop skills of foreign languages.
- Grammatical Sensitivity which is the ability to acquire knowledge and awareness of the distinctive features and structural principles of a native or a foreign language that allows for the construction of words (morphology) and sentences (syntax).
- Listening Ability, which is the ability to listen and understand the meaning of oral communications (spoken words, phrases, sentences, and paragraphs).
- Communication Ability, which is the ability to speak in real life situations (e.g., lecture, conference and group participation) in a manner that transmits ideas, thoughts, or feelings to one or more individuals.

While many people claim that their intelligence seems to decline as they age, research suggests that while fluid intelligence begins to decrease after adolescence, crystallized intelligence continues to increase throughout adulthood. Fluid and crystallized intelligence are complementary in that some learning tasks can be mastered mainly by exercising either fluid or crystallized intelligence.

Both types of intelligence increase throughout childhood and adolescence. The fluid intelligence is maximized in adolescence and begins to decline progressively around age 40. Crystallized intelligence does not decline, but continues to grow throughout adulthood.

2.10 General Visualization (Gv) and General Auditory (Ga) Perceptual Factors

2.10.1 General Visualization

The question of how the brain sees, recognizes and understands objects is one of the most intriguing in neuroscience. The use of visualization has often been cited as a powerful problem representation process for solving problems within the field of mathematics. It has been argued that the use of visual images can be an important help for all sorts of problems, including problems in which nothing geometric is evident. Visual imagery, according to Owens and Clements (1998), has a role in establishing the meaning of a problem, channeling problem-solving approaches, and spatial manipulations. Spatial visualization is the ability to mentally manipulate 2-dimentional and 3-dimentional figures. Visual imagery refers to the ability to form mental representations of the appearance of objects and to manipulate these representations in the mind (Kosslyn, 1995). Visual imagery tests and learning can enhance mathematical problem solving. Neurologists reported that the spatial ability factor was one of the main factors significantly affecting mathematical performance. On the basis of such reports, It is concluded that individuals can be classified into three groups according to how they process mathematical information. The first group consists of verbalizers, who prefer verballogical rather than imagery modes when attempting to solve problems; the second group, visualizers, involves those who prefer to use visual imagery; and the third group, mixers, contains individuals who have no tendency one way or the other.

Visual imagery refers to a representation of the visual appearance of an object, such as its shape, colour, or brightness. Spatial imagery refers to a representation of the spatial relationships between parts of an object and the location of objects in space or their movement. Further, spatial imagery is not limited to the visual modality (i.e., one could have an auditory or haptic spatial image).

Visual-Spatial Abilities include:

- Visualization which is the ability to apprehend a spatial form, object, or scene and match it with another spatial object, form, or scene with the requirement to rotate it (one or more times) in two or three dimensions. Requires the ability to mentally imagine, manipulate or transform objects or visual patterns (without regard to speed of responding) and to see (predict) how they would appear under altered conditions (e.g., parts are moved or rearranged). The stimulus representation could be visual-graphic, oral, etc., and the response mode could be verbal, pointing, drawing, etc.
- Perceptual speed, fluency (figural, expressional, ideational, and word), reasoning (general and logical), spatial (orientation and scanning) and verbal comprehension.
- Flexibility of Closure which is the ability to identify a visual figure or pattern embedded in a complex distracting or disguised visual pattern or array, when knowing in advance what the pattern is. Recognition of, yet the ability to ignore, distracting background stimuli is part of the ability.
- Length Estimation, which is the ability to accurately estimate or compare visual lengths or distances without the aid of measurement instruments.
- Perceptual Illusions, which is the ability to resist the illusory perceptual aspects of geometric figures (i.e., not forming a mistaken perception in response to some characteristic of the stimuli).

- Imagery, which is the ability to mentally depict (encode) and/or manipulate an object, idea, event or impression (that is not present) in the form of an abstract spatial form.

2.10.2 General Auditory

While there are many similarities between our perception of visual and auditory information, there are a few fundamental differences. One key difference is that auditory events are dynamic-they're experienced over time. Researchers refer to this fact as auditory streaming. An auditory stream is the perceptual unit of sound, much like the object is the perceptual unit of vision. For example, an auditory stream could be the low hum of a heating fan or a quiet group of people conversing in the background. Our ability to separate auditory streams into appropriate regions or groups is referred to as auditory scene analysis. Given an auditory scene, auditory objects can be integrated to form a perceptual stream in one of two ways: sequentially or simultaneously. Sequential integration means that auditory events are connected over time, while simultaneous integration means that auditory events share spectral qualities, such as frequency, at the same moment in time.

The brain receives and transmits auditory frequencies through a continuous change in the neurosystem of the brain, Figure (2.12).

Figure (2.12): Change of neurosystem in the brain according to the frequency of speech

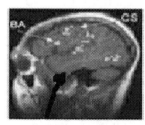

silently saying
animal name

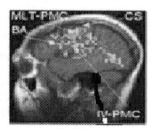

saying animal
name

http://www.neurolearning.com/images/littlevoice.jpg

It has been demonstrated that acoustic, phonological, semantic and syntactic context can all shift the perceived identity of a speech sound (Borsky, Tuller, & Shapiro, 1998). As a result, two identical acoustic segments can be labeled as different phonemes and two different acoustic segments can be labeled with identical phonemic labels.

Abilities of auditory include:

- Phonetic Coding, which is the ability to code, process, and be sensitive to nuances in phonemic information (speech sounds) in the short-term memory. This includes the ability to identify, isolate, blend, or transform sounds of speech. It is frequently referred to as phonological or phonemic awareness.
- Speech Sound Discrimination which is the ability to detect and discriminate differences in phonemes or speech sounds under conditions of little or no distraction or distortion.
- Maintaining and Judging Rhythm, which is the ability to recognize and maintain a musical beat.
- Sound-Frequency and Sound-Intensity discrimination which is the ability to discriminate frequency and amplitude of sound pattern.
- Musical Discrimination and Judging Rhythm which is the ability to discriminate and judge musical rhythm and beat.
- Resistance to Auditory Stimulus Distortion which is the ability to overcome the effects of distortion or distraction when listening to and understanding speech and language.
- Temporal Tracking which is the ability to mentally track auditory temporal (sequential) events so as to be able to count, anticipate or rearrange them (e.g., reorder a set of musical tones).

2.11 Cognitive Processes

Cognitive processes (also known as mental function) which indicate such functions or processes as perception, introspection, memory, imagination, conception, belief, reasoning, volition, and emotion. So cognitive processes are those involved in the acquisition, processing and use of knowledge and information.

Cognitive processing speed is your thinking speed; it is a measure for the responsiveness of your brain, and correlates very well with brain health and age, but also with intelligence and education. Quick speed of processing is important for decision making in our daily lives particularly in tasks like driving where important decisions must be made quickly.

2.11.1 Classification of Cognitive Processes

The cognitive processes are divided into three domains: the cognitive, the affective, and the psychomotor. The cognitive is commonly referred to as Bloom's Taxonomy of the Cognitive Domain (Bloom et al., 1956). The Bloom's taxonomy was revised by Anderson and Krathwohl (2001) to fit the focused modern education objectives with more focused outcomes. There were no big changes in the bloom's taxonomy, except that the names of the levels were changed from nouns to active verbs, the orders of the highest two levels were reversed, and the lowest-order level (Knowledge) was changed to remembering. The table (2.2) below shows the bloom's taxonomy:

Table (2.2): Bloom's taxonomy

LEVEL	DEFINITION	SAMPLE VERBS	SAMPLE BEHAVIORS
KNOWLEDGE	Student recalls, identifies or recognizes information, ideas, and principles in the approximate form in which they were learned.	Write List Label Name State Define	The student will define the 6 levels of Bloom's taxonomy of the cognitive domain.
COMPREHENSION	Student translates, comprehends, extrapolates or interprets information based on prior learning.	Explain Summarize Paraphrase Describe Illustrate	The student will explain the purpose of Bloom's taxonomy of the cognitive domain.
APPLICATION	Student selects, relates, transfers, associates and uses data and principles to complete a problem or task with a minimum of direction.	Use Compute Solve Demonstrate Apply Construct	The student will write an instructional objective for each level of Bloom's taxonomy.
ANALYSIS	Student distinguishes, classifies, discriminates and relates the assumptions, hypotheses, evidence, or structure of a statement or question.	Analyze Categorize Compare Contrast Separate	The student will compare and contrast the cognitive and affective domains.

LEVEL	DEFINITION	SAMPLE VERBS	SAMPLE BEHAVIORS
SYNTHESIS	Student originates, integrates, constitutes , formulates, specifies and combines ideas into a product, plan or proposal that is new to him or her.	Create Design Hypothesize Invent Develop	The student will design a classification scheme for writing educational objectives that combines the cognitive, affective, and psychomotor domains.
EVALUATION	Student appraises, assesses, or critiques on a basis of specific standards and criteria.	Judge Recommend Critique Justify	The student will judge the effectiveness of writing objectives using Bloom's taxonomy.

So, Bloom's taxonomy of the cognitive domain classifies the cognitive process into six dynamic levels of increasing complexity, from knowledge as the baseline, through comprehension, application, analysis, and synthesis, to evaluation as the highest level. This classification, if appreciated by medical teachers and students and correctly applied, should make meta-cognition of the diagnostic process routine. A correct diagnosis will in most cases lead to appropriate treatment and lead to describe clinical information including functional or anatomic, pathologic as well as etiologic conclusion.

Usually a condition affecting one body system can produce structural or functional abnormality in different parts of the body. One therefore needs to acquire knowledge of the pathological changes and the clinical manifestations of various diseases in particular systems. This will enable one to translate the symptoms and signs to a particular system and from their character one may begin to associate them to particular types of pathological processes. With knowledge of the features of different pathological processes one may distinguish a particular process from others, and coupled to one's knowledge of epidemiology of diseases one may begin to formulate a diagnosis. So, at every point in the processing of information one moves from one level to the other in increasing order of complexity, at times going back and forward. Without appropriate knowledge, it is difficult to proceed logically to the next level of cognition. A lower level can always be used to facilitate a higher level of cognition,[http://www.med-ed-online.org/f0000007.htm]

2.12 Relationship of Intelligence

Your intellectual intelligence and emotional intelligence, together with your body intelligence comprise essential parts of your wisdom and your judgment. You can use them to gauge your interpersonal maturity.

Relationship of intelligence is a hologram of your life. You cannot hide your self-awareness, your maturity, your self-control, your commitment, your ethics and your integrity. In every relationship you will show how well you can inspire and lead, monitor your ability for success, control strong emotions and impulses, communicate, initiate change, follow through and solve problems.

2.12.1 Cognitive Style

Cognitive style or "thinking style" is a term used in cognitive psychology to describe the way individuals think, perceive and remember information, or their preferred approach to using such information to solve problems. Cognitive style differs from cognitive ability (or level), the latter being measured by aptitude tests or so-called intelligence tests. Controversy exists over the exact meaning of the term cognitive style and also as to whether it is a single or multiple dimension of human personality. However, it remains a key concept in the areas of education and management.

Considerable confusion appears in the literature concerning the terms cognitive style and learning style. Numerous authors use the terms interchangeably.

Psychologists introduced several models to identify and measure the cognitive style including:

Cognitive Style Analysis (CSA), which is a compiled computer-presented test that measures individuals' position on two orthogonal dimensions:

- Wholist-Analytic (W-A), which reflects how individuals organise and structure information. Wholists retain a global or overall view of information. Wholist-analytic cognitive style can be defined as the tendency for individuals to process information either as an integrated whole or in discrete parts of that whole. In practical terms, analytics are able to apprehend ideas or concepts in parts, but have difficulty integrating such ideas into complete wholes. However, wholists are able to view ideas as complete wholes, but are unable to separate these ideas into discrete parts as shown below:

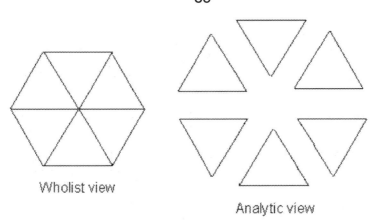

Wholist view

Analytic view

- Verbal-Imagery (V-I) which describes individuals' mode of information representation in memory during thinking. The verbaliser-imager cognitive style can be defined quite simply as an individual's tendency to process information either in words or in images. Verbalisers are superior at working with verbal information, (Riding and Mathias, 1991; Riding and Watts, 1997) whereas imagers are better at working with visual and spatial information.

- Time-Response, which is the ability of the respondent to answer quickly. It is a mixture of cognitive style and cognitive ability. The Time- Response is an unreliable method of testing the cognitive style.

- Another area where individuals show differences in their abilities to discriminate events or visual, auditory, or tactile cues from their surrounding environments is known as field-dependence/field-independence.The Field dependence-independence model which identifies an individual's perceptive behaviour while distinguishing object figures from the content field in which they are set. Two similar instruments to do this were produced, the Embedded Figures Test (EFT) and the Group Embedded Figures test (GEFT). In both cases, the content field is a distracting or confusing background. These instruments are designed to distinguish field-independent from field-dependent cognitive types; a rating which is claimed to be value-neutral. Field-independent people tend to be more autonomous when it comes to the development of restructuring skills; that is, those skills required during technical tasks with which the individual is not necessarily familiar. They are, however, less autonomous in the development of interpersonal skills. The EFT and GEFT continue to enjoy support and usage in research and practice. However, they too, are criticized by scholars as containing an element of ability and so may not measure cognitive style alone.

- Convergent-Divergent thinkers in which convergent thinkers are good at accumulating material from a variety of sources relevant to a problem's solution, and divergent thinkers proceed more creatively and subjectively in their approach to problem-solving (Hudson 1967). Converger-Diverger constructs attempt to measure the processing rather than the acquisition of information by an individual. It aims to differentiate convergent from divergent thinkers; the former being persons who think rationally and logically, while the latter tend to be more flexible and to base reasoning more on heuristic evidence. Convergent thinking, in which the person is good at bringing material from a variety of sources to solve a problem in such a way as to produce the "correct" answer. This kind of thinking is particularly appropriate in science, math and technology. Because of the need for consistency and reliability, this is really the only form of thinking, which standardized intelligence tests (and even national exams) can test. With divergent thinking the student's skill is in broadly creative elaboration of ideas prompted by a stimulus, and is more suited to artistic pursuits and study in the humanities.

- Cognitive Complexity theories attempt to identify individuals who are more complex in their approach to problem-solving against those who are simpler. Cognitive complexity is a psychological characteristic or psychological variable that indicates how complex or simple the frame and perceptual skills of a person are. A person who is measured high on cognitive complexity tends to perceive nuances and subtle differences which a person with a lower measure, indicating a less complex cognitive structure for the task or activity, does not.

- Holists –Serialists in which the Holists gather information randomly within a framework, while serialists approach problem solving step-wise, proceeding from the known to the unknown. See the difference in the illustration below:

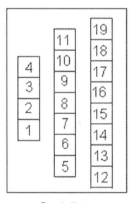 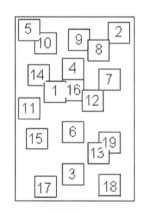

Serialists   Holists

- Hemispherical Lateralisation Concept, commonly called left-brain/right-brain theory (the human brain is a paired organ; it is composed of two halves called cerebral hemispheres that look pretty much alike), posits that the left hemisphere of the brain controls verbal, logical and analytical operations, while the right hemisphere controls nonverbal (responding to touch and music), holistic, intuitive, sensory and pictorial activities. Cognitive style is thus claimed to be a single dimension on a scale from extreme left-brain to extreme right-brain types, depending on which associated behaviour dominates in the individual, and by how much.

- The Allinson-Hayes Cognitive Style Index (CSI) has features of the left-brain / right-brain theory. The CSI contains 38 items, each rated using a 3-point scale (true; uncertain; false). Some scholars have questioned the CSI's construct validity on the grounds of theoretical and methodological limitations associated with its development. It is also noteworthy that this measure of cognitive style is both gender-sensitive and culture-sensitive. While it is entirely plausible that cognitive style is related to these social factors, it does complicate some educational and management issues. It suggests, for instance, that a given student is best taught by a person of a certain sex or culture; or that only persons of certain cultures can work harmoniously together in teams. The CSI is intended primarily for use in business.

2.12.2 Conceptual Tempo or Cognitive Tempo

Conceptual tempo or cognitive tempo is an area of individual difference in the speed of information processing. It can be regarded as an element of cognitive style as some people have a quicker cognitive processing speed over a whole range of cognitive processes. While cognitive style was not found to be related to strategies employed during problem solving, conceptual tempo influenced both time spent and quantity and quality of questions asked.

2.12.3 Perceptual Speed

Perceptual speed is the ability to quickly and accurately compare letters, numbers, objects, pictures, or patterns. In tests of perceptual speed the things to be compared may be presented at the same time or one after the other. Candidates may also be asked to compare a presented object with a remembered object.

Many intelligence tests, such as aptitude tests, personality tests, EQ and IQ tests, measure candidates' perceptual speed, including the Thomas International General Intelligence Assessment. Statistically, women tend to outperform men in terms of perceptual speed.

2.12.4 Reaction Time

Reaction time is the time from the onset of a stimulus until the emission of an organism response. There are two types of reaction time: Simple reaction time which is the time it takes to react to stimuli, and complex reaction time which is the latency between a variable stimulus and a respectively variable response. Factors affecting reaction time can be one or more of the followings:

- Recognition
- Number and type of stimuli

- Stimulus intensity
- Choice
- Disease, such as Parkinson's, Huntington's, chicken pox, etc
- Distractions
- Fatigue
- Gender
- Age
- Race
- Hand and finger tremors
- Practice and error
- Vision
- Right vs. left hand
- Breathing cycle
- Heart and lung disease
- Stimulant and antidepressant drugs

2.12.5 Words per Minute

Words per minute, (commonly abbreviated wpm) are a measure of input or output speed and in psychology can be related to cognitive processing speed and perceptual speed. Alphanumeric entry, handwriting, reading and comprehension, and speech and listening are also related to the cognitive processing speed.

2.13 Parieto-Frontal Integration Theory (P-FIT) of Intelligence

Studies from functional (i.e., functional magnetic resonance imaging, positron emission tomography) and structural (i.e., magnetic resonance spectroscopy, diffusion tensor imaging, voxel-based morphometry) neuroimaging paradigms reveal striking findings suggesting that variations in a distributed network in the brain predict individual differences found on intelligence and reasoning tasks. Such a network is called the Parieto-Frontal Integration Theory (P-FIT), which stated that that human intelligence arises from a distributed and integrated neural network comprising brain regions in the frontal and parietal lobes, Richard Haier & Rex Jung (July 26, 2007). "The Parieto-Frontal Integration Theory (P-FIT) of intelligence: Converging neuroimaging evidence", Cambridge University Press.

The P-FIT model includes, by Brodmann areas (BAs): the dorsolateral prefrontal cortex (BAs 6, 9, 10, 45, 46, 47), the inferior (BAs 39, 40) and superior (BA 7) parietal lobule, the anterior cingulate (BA 32), and regions within the temporal (BAs 21, 37) and occipital (BAs 18, 19) lobes. White matter regions (i.e., arcuate fasciculus) are also implicated, Figure (2.13):

Figure (2.13): Areas of the P-FIT

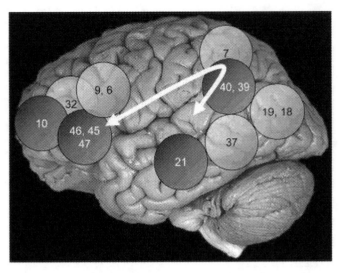

http://www.scientificblogging.com/news_account/parie
to_frontal_integration_theory_p_fit_a_neural_basis_of
_intelligence

In conclusion, the P-FIT is still disputed by many psychologists and neuroscientists arguing that cognitive and neural structures and regions mediating performance change as skill increases. Therefore, the structures highlighted by Parieto-Frontal Integration Theory are unlikely to account for individual differences in skilled cognitive achievement in everyday life.

2.14 Some Philosophical Problems from the Standpoint of Intelligence

There is no agreement on what intelligence really is. Different branches of science use different definitions. In this context, it is problematic to claim that the intelligence quotient is a measure of intelligence.

- Intelligence may be composed of many different aspects. Some scientists think it is problematic that these different aspects can be combined into one "measurement".
- The first tests were done on children in school, to determine which children would likely need more attention. This is completely different from measuring "intelligence". A child that needs more help in school is not necessarily less intelligent; it might simply come from a different background. For example, Early in the 20th century, IQ tests were used to screen foreign immigrants to the United States; roughly 80% of Eastern European immigrants tested during the World War I era were declared "feeble-minded," even though the tests discriminated against them in terms of language skills and cultural knowledge of the United States.
- Some tests require that those tested come from a certain cultural background. People outside the culture will test badly, but they may not

be less intelligent. For example, in some cultures, girls were discouraged from pursuing mathematics by teachers, peers, and parents.

2.15 Artificial Neural Networks (Neural Intelligence)

An artificial neural network is, in essence, an attempt to simulate the brain and to mimic the properties of biological neurons. The Neural network theory revolves around the idea that certain key properties of biological neurons can be extracted and applied to simulations, thus creating a simulated (and very much simplified) brain. The first important thing to understand them is that the components of an artificial neural network are an attempt to recreate the computing potential of the brain. The second important thing to understand, however, is that no one has ever claimed to simulate anything as complex as an actual brain. The real, biological nervous system is highly complex and includes some features that may seem superfluous based on an understanding of artificial networks, as the human brain is estimated to have something on the order of ten to a hundred billion neurons. A typical artificial neural network (ANN) is not likely to have more than 1,000 artificial neurons. Artificial intelligence and cognitive modeling try to simulate some properties of neural networks. While similar in their techniques, the former has the aim of solving particular tasks, while the latter aims to build mathematical models of biological neural systems.

In the artificial intelligence field, artificial neural networks have been applied successfully to speech recognition, image analysis and adaptive control, in order to construct software agents (in computer and video games) or autonomous robots. Most of the currently employed artificial neural networks for artificial intelligence are based on statistical estimation, optimization and control theory.

The cognitive modelling field involves the physical or mathematical modeling of the behaviour of neural systems; ranging from the individual neural level (e.g. modeling the spike response curves of neurons to a stimulus), through the neural cluster level (e.g. modelling the release and effects of dopamine in the basal ganglia) to the complete organism (e.g. behavioural modeling of the organism's response to stimuli). Artificial intelligence, cognitive modelling, and neural networks are information processing paradigms inspired by the way biological neural systems process data.

In theory, an artificial neuron (often called a 'node') captures all the important elements of a biological one. Nodes are connected to each other and the strength of that connection is normally given a numeric value between -1.0 for maximum inhibition, to +1.0 for maximum excitation. All values between the two are acceptable, with higher magnitude values indicating stronger connection strength. The transfer function in artificial neurons, whether in a computer simulation or actual microchips wired together, is typically built right into the nodes' design. The transfer functions of the artificial neural network are mathematical formulas and can not represent the actual biological behaviour of the neurosystem.

2.16 Brian-Signal Simualtion

The human brain is very complex in sending and receiving signals of actions and movements. Each signal has a start and stop point. The researchers explored the role of brain circuits located in the basal ganglia, and found that dopamine-producing neurons project into the striatum and terminate at the substantia nigra. Substantia nigra is located in the basal ganglia, and both play an important role in the initiation and termination of newly learnt behavioural sequences. Substantia nigra means "black substance" in Latin, as parts of the substantia nigra appear darker than neighboring areas due to high levels of melanin in the neurotransmitters (dopamine). The basal ganglia contains the striatum, globus pallidus and subthalamus nucleus.

Rui Costa and Xin Jin (National Institutes of Health,USA) show that when mice are learning to perform a particular behavioural sequence there is a specific neural activity represented in electrical signals that emerges in those brain circuits and signals the initiation and termination steps. Interestingly these are the circuits that degenerate in patients suffering from Parkinson's and Huntington's diseases, who also display impairments both in sequence learning, and in the initiation and termination of voluntary movements.

Xin Jun adds: "This start/stop activity appears during learning and disrupting it genetically severely impairs the learning of new action sequences. These findings may provide a possible insight into the mechanism underlying the sequence learning and execution impairments observed in Parkinson's and Huntington's patients who have lost basal ganglia neurons which may be important in generating initiation and termination activity in their brain".

The network of brain circuitries is too complex to describe, but molecular biology and computing methods have improved to the point that the National Institutes of Health have announced a $30 million plan to map the human "internal connection and signal pathways".

University of Southern California neuroscientists Richard H. Thompson and Larry W. Swanson used the method to trace circuits running through a "hedonic hot spot" related to food enjoyment. The circuits showed up as patterns of circular loops, suggesting that at least in this part of the rat brain, the wiring diagram looks like a distributed network of the Internet.

"We started in one place and looked at the connections. It led into a very complicated series of loops and circuits. It's not an organizational chart. There's no top and bottom to it," Swanson said.

The author of this book (Dr. Amin Elsersawi) believes that incoming and outgoing signals from basal ganglia to each part of the brain are similar to a spider web of multi-feed back loop. If you knock out any single "thread" of the web the rest of it works. He agreed with Swanson when she said, "There are usually alternate pathways through the nervous system. It's very hard to say that any one part is absolutely essential."

Dr. Elsersawi suggested an imaginative electrical diagram representing the brain network as depicted in Figure (2.14).

Figure (2.14): Signal pathways in the brain

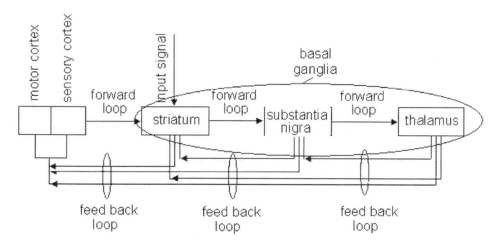

2.16.1 Brain Maps and the Secret to Intelligence (Higher IQ)

1. New research suggests that the layer of insulation coating neural wiring in the brain (myelin) plays a critical role in determining intelligence. In addition, the quality of this insulation appears to be largely genetically determined, providing further support for the idea that IQ is partly inherited. The neural wires that transmit electrical messages from cell to cell in the brain are coated with a fatty layer called myelin. Much like the insulation on an electrical wire, myelin stops currents from dissipating out of the wire and keeps the voltage level constant with which messages travel through the brain--the higher quality the myelin, the faster the messages travel. These myelin-coated tracts make up the brain's white matter, while the bodies of neural cells are called grey matter.

2. Does white matter play a key role in intelligence? The size of the corpus callosum, the thick tracts of white matter connecting the two hemispheres of the brain, is about 95 percent genetic. And about 85 percent of the white-matter variation in the parietal lobes, which are involved in logic and visual-spatial skills, can be attributed to genetics (according to the neurologist Thompson at UCLA). But only about 45 percent of the variation in the temporal lobes, which play a central role in learning and

memory, appears to be inherited. Thus, it's likely that the quality of white matter is at least partly genetically determined and, therefore, difficult to change.

3. Researchers have yet to find a simple neural explanation for intelligence. In 2001, Thompson showed that it is correlated with volume in the frontal cortex, a result consistent with a number of studies that have linked intelligence to overall brain size. But size is a crude measure: while larger brains may be smarter on average, it's not clear if that's because they have more nerve cells, more connections between cells, more of the fibers that carry neural signals, or less nodes that connect messages.

4. Scientists at UC Santa Barbara have made a major discovery in how the brain encodes memories. The team of scientists is the first to uncover a central process in encoding memories that occurs at the level of the synapse, where neurons connect with each other. Strengthening synapses is a very important part of learning and memory.

2.17 Alternative Approaches to Cognitive Abilities

To this day, how exactly to define intelligence is still debated. There are, however, two major schools of thought on its nature and properties. The two opposing theories of intelligence are the one general intelligence school of thought (Charles Spearman) and the multiple intelligences school of thought (Howard Gardner). The general intelligence proponents believe that there is one factor from which all intelligence is derived; the multiple intelligences proponents believe that there are different kinds of intelligence. Each theory has merit and evidence to support its claims.

2.17.1 Spearman's Intelligence

Charles Spearman has a "Model of Intelligence Theory". In the early 1900's Charles Spearman made an observation that has since continued to influence many of today's theories of intelligence. In this observation, Spearman noted that all tests of mental ability are positively correlated. Spearman discovered that people who score high on IQ or mental ability tests usually scored higher on other types of tests, and people that scored lower generally had lower scores on other tests.

Furthermore, there is a very high correlation between IQ and very simple cognitive tasks, which supports the theory of one general intelligence (Eysenck, 1982).

Spearman proposed the idea that intelligent behavior is generated by a single, unitary quality within the human mind or brain. Spearman derived this theoretical entity, called the general factor, or simply g, through a new statistical technique that analyzed the correlations among a set of variables. This technique, called factor analysis, demonstrated that scores on all mental tests are positively correlated; this offered compelling evidence that all intelligent behavior is derived from one metaphorical pool of mental energy. Although proponents of multiple intelligence theory reject this interpretation, factor analysis remains one of the most important tools in 21st century intelligence research.

Charles Spearman stated that two factors could directly affect an individual's score on mental tests. He called these factors:

- The general intelligence or the general factor. This factor represents what all mental tests have in common. Scores resulted from these tests are positively correlated, as he believes that all of these tests based on the general factor. When he examined the results of these different tests, he found that there was a positive correlation between the tests for a given individual. In other words, if a certain person performed well on a test of verbal abilities, then that same person also performed well on another test of another cognitive ability, (for instance, a mathematics test). Spearman and his followers believe that intelligence can be defined by a single factor. Whether that single factor is termed positive manifold, neural processing speed, or g, the complexities of the human mind and its processes can be reduced to a single factor, defined as intelligence.
- Specific factor which is related to whatever unique abilities a particular test required. This factor is differed from test to test. Spearman and his supporters place much more importance on general intelligence than on the specific factor.

2.17.2 Seven Intelligences by Howard Gardner (or Multiple Intelligence)

Howard Gardner proposed a theory that was intended to broaden the traditional definition of intelligence. He felt that the concept of intelligence, as it was previously defined, did not truly capture all of the ways that humans can excel. Gardner argued that human beings do not have underlying general intelligence, but rather that we have multiple intelligence's and that they are each part of an independent system in the brain. One example of this is the studies he did on brain damaged people who had lost one ability (such as spatial thinking) but managed to retain another (such as language or motor functions). He believed that because two separate abilities could operate independently of one another, this suggested that the existence of separate intelligences was not that far fetched. He also suggested that evidence of multiple intelligences could also be the various idiot savants or prodigies. He believed that the presences of high level or extraordinary abilities in the absence of other abilities more than solidified his theory.

He also proposed that there where seven intelligences. The first intelligence was Linguistic Intelligence, which involves aptitude with speech and language. The next was Logical-Mathematical Intelligence; this involves the ability to reason abstractly and solve mathematical and logical problems. The third is Spatial

Intelligence; this is used to perceive visual and spatial information and to conceptualize the world in tasks, such as navigation or art. Musical Intelligence is the fourth ability. This is the ability to perform, read, write or decipher music. The fifth ability is Bodily- Kinesthetic intelligence; this is the ability that people use in activities such as sports, dancing, and so on. Interpersonal intelligence is the sixth which is the ability to understand others while Intrapersonal intelligence is the ability to understand ones self. Finally, we have "Naturalist intelligence, which is the ability to recognize and classify plant life" (Encarta.msn, 2006).

2.18 Genotype Intelligence

Fluid intelligence (g_f) is a major dimension of individual differences in cognitive function. A strong predictor of achievement in educational and other domains, at a mechanistic level, g_f has been linked to processes ranging from cognitive flexibility and strategy development to manipulation of stored mental representations and attention control. In particular, the inhibition of interference (Kane et al.2005). Fluid intelligence (g_f) is the factor that defines the phenotype for intellectual functioning. It is no overstatement to say that g_f and g_c are undoubtedly the most important psychological constructs discovered in this century. It would seem that g is the perfect phenotypic definition of intelligence.

It has been shown that genetic factors influence the fluid intelligence (g_f). It is also affected by environmental factors. Genetic influences account for 40% or more of the variance in g_f scores (Gray and Thompson 2004). This has been verified when measured by the blood oxygen level–dependent (BOLD) signal in functional magnetic resonance imaging (fMRI).

The behavioural and Clinical Neuroscience Institute, University of Cambridge, UK reported that fluid intelligence (g_f) influences performance across many cognitive domains. It is affected by both genetic and environmental factors. Tasks tapping g_f activate a network of brain regions including the lateral prefrontal cortex (LPFC), the presupplementary motor area/anterior cingulate cortex (pre-SMA/ACC), and the intraparietal sulcus (IPS). In line with the "intermediate phenotype*" approach, we assessed effects of a polymorphism (val[158]met**) in the catechol-O-methyltransferase (COMT) gene on activity within this network and on actual task performance during spatial and verbal g_f tasks. COMT regulates catecholaminergic signaling in prefrontal cortex. The val[158] allele is associated with higher COMT activity than the met[158] allele.

* A phenotype is any observable characteristic or trait of an organism: such as its morpology, development, biochemical or physiological properties, behaviour, and products of behavior. Phenotypes result from the expression of an organism's genes as well as the influence of environmental factors and the interactions between the two; see Biochemistry of Aging by the author.

** Catechol-O-methyltransferase (COMT) inactivates dopamine, epinephrine and norepinephrine in the nervous system. A common functional polymorphism (Val158Met) leads to a three- to-four-fold variation in the COMT enzyme activity, the Met form displaying lower enzymatic activity.

2.19 Reflective Intelligence

A battle has raged among psychologists for several decades about the true nature of intelligence. They debate about three distinct types of intelligence:

- Neural intelligence – we are born with neural intelligence and it never changes, although there are nutritional and maturational effects. Most psychologists consider intelligence a matter of the efficiency of the nervous system, genetically determined, not very subject to environmental influence, and adequately measured by IQ tests.
- Experiential intelligence - we gain experiential intelligence through experience in a specific area, such as playing music.
- Reflective intelligence – we gain reflective intelligence by being aware of our thinking patterns and the way we can change these patterns. Still many psychologists emphasize how good thinking depends on good mental management (knowing what questions to ask yourself and what answers to reply), using problem-solving strategies, monitoring and striving to direct and improve your own thinking. Reflective intelligence consists of the planning, strategy, and attitude that help us to think and act more effectively. For example, whether chess or bridge or football, players who have thought through a strategy for advancing their side will generally do better than those who simply play defensively and react to the other side's moves. Reflective intelligence offers the best target of opportunity for education because reflective intelligence is the most learnable of the three.

2.20 Intelligence Paradox

Race and intelligence research investigates differences in the distributions of cognitive skill measurements among human racial groups. Much of the debate in this area centers around questions of how intelligence is measured, and the relative influence that genetics and environment have on both intelligence and attempts to measure it. Debates in popular science and academic research over the possible relation between racial divisions and differences in intelligence originally began as a comparison of African Americans and Caucasians in the United States, but were later extended to other ethno-racial groups and regions of the world. In the US, intelligent quotient (IQ) test scores show statistical differences, with the average score of the African American population being lower - and that of the Asian American population being higher - than that of the White American population (based on the self-identification of those tested).

Prof. James Flynn (along with most neuroanthropological philosophers) famously pointed out to people outside the standardized testing industry that IQ tests had to be periodically recalibrated because average IQ scores in industrialized countries steadily inflated, suggesting either that people were growing smarter or something else was up with these tests. Flynn gathered tests from Europe, North America and Asia, around thirty countries in all, and discovered that, for as far back as we had data in any case, average IQ test scores had risen about 3 points per decade and in some cases more. Only recently, in some Scandanavian countries, do the gains appear to be levelling off (see, for example, Sundet 2004; Teasdale and Owen 2005).

Flynn noticed that there is a steady rise from one generation to the next in average scores on IQ tests over time. Flynn noted that this effect apparently contradicts some fundamental beliefs about IQ long held by intelligence researchers, and categorizes these contradictions into four seeming "paradoxes":

1. The factor analysis paradox - past research has shown evidence for a single factor, "g" or general intelligence, underlying IQ (Freud's claim that memory is permanent and unchangeable). However, the Flynn Effect happens to different degrees in the sub-tests of the WISC intelligence test, suggesting that intelligence as measured by IQ tests is multidimensional. Flynn poses this as: "how can intelligence be both one and many at the same time"?
2. The intelligence paradox - the Flynn effect shows significant improvements in IQ over a short time scale, yet we do not notice in everyday life that young people are significantly smarter than their parents or grandparents. For example, modern families have far less children than families from only 50 years in the past, and consequently parents are able to spend more time with each child, thus stimulating their cognitive abilities at an early age, which is obviously beneficial to overall intelligence as well as psychological well-being.
3. The mental retardation paradox - the IQ level commonly associated with mental retardation is 75. If the Flynn effect is extrapolated back to 1900 the mean IQ would be somewhere between 50 and 70 - that is If IQ gains are in any sense real, we are driven to the absurd conclusion that a majority of our ancestors were mentally retarded.
4. The identical twins paradox - past IQ research has shown a close relation between the IQs of identical twins reared separately; a fact used as evidence for a genetic basis for differences in IQ. The rapid changes in IQ shown by the Flynn effect suggests, conversely, that environmental factors have a greater influence on IQ than genes. If the genes dominate individual differences then how can environment be both so feeble and so potent in twin studies?

The IQ Paradox has been a major topic for debate and attention for some time; however the cause of such a dramatic rise in IQ over the past century has not been fully explained.

2.21 Bodily Intelligence

Bodily/Kinesthetic intelligence (body smart) is the capacity to use your whole body or parts of your body (your hands, your fingers, your arms), to solve a problem, make something, or put on some kind of production. Each person possesses a certain control of his or her movements, balance, agility and grace. The most evident examples are people in athletics or the performing arts, particularly when dancing or acting. Example: Tiger Woods and Michael Jordan. Some people, however, argue that physical control does not constitute a designation as a form of intelligence. But the work of Gardner and other MI researchers maintains that bodily-kinesthetic ability does indeed deserve such recognition.

2.22 Quantitive Reasoning

Often, quantitative reasoning (QR) is assumed to be synonymous with mathematics, and, indeed, the two are inextricably linked.
Basic quantitative skills include arithmetic, units, scientific notation, linear equation with one or two unknowns, linear and quadric equations, exponents and logarithms, nonlinear representation and slope as instantaneous rate of change.

Yet there are differences, one of which is that while mathematics is primarily a discipline, QR is a skill, one with practical applications. A mathematician might take joy in abstraction, but the well-educated citizen can apply QR skills to daily contexts. For instance, understanding the power of compound interest or the uses and abuses of percentages; using fundamental statistical analysis to gauge the accuracy of a statistical study; or applying the principles of logic and rhetoric to real world arguments.

2.23 General Nonverbal Ability

Jack A. Naglieri, PhD stated that nonverbal ability tests (which are called Naglieri Nonverbal Ability Tests – NNAT) provide a valid way to measure general ability for all children, regardless their cultural and linguistic background. Researchers have found that they identify similar proportions of black, white, and Hispanic children as gifted. This suggests that the problem of under representation of minority children in classes for the gifted may be addressed by using such tests. Using nonverbal tests of general ability gives all children an equal opportunity to succeed. In summary, nonverbal tests provide a valid way to measure general ability for all children. It is primarily used to identify the following:

- Gifted students for whom English is a second language

- Children whose school performance may be considered poor because of limited English proficiency
- To identify at-risk students whose low nonverbal reasoning ability may indicate potential academic problems
- To identify children who may have a learning disability and require further diagnostic testing

Time: Students are given 30 minutes to complete 39 multiple choice questions. Total test time is approximately 40 minutes.
Segments: Pattern Completion, Reasoning by Analogy, Serial Reasoning, and Spatial Visualization.

Examples of IQ Tests

1- Numerical reasoning test

What is the missing number? 1   4   9   ?   25

2- Sequences (logical reasoning) test

What is the missing letter in this series (considering the alphabet series)
a   c   e   ?   i

3- Computer programming aptitude test

John thought of a number, added 8, multiplied by 4, took away 5 and divided by 7 to give an answer of 5.
What was her starting number?

4- Vocabulary test

What is the meaning of paradox? Self-contradictory statement, weed killer or type of ghost.

2.24 Cognitive Elite

After the Second World War, macro sociological changes, such as rapid industrialization, federal institution expansion, academic and research centers spreading out, and the construction of new cities and suburban areas were reflected in an increase of American's educational skills and achievement, occupational mobility and wealth. These changes symbolized the nation's emergence as an affluent country. This is the cognitive elite of a society.

The cognitive elite of a society, according to some social science researchers, are those having higher intelligence levels and are thus better prospects for success in life. The development of a cognitive elite during the 20th century is presented in the book the Bell Curve written by Richard J. Herrnstein and Charles Murray and published by Free Press Paperbacks in 1996. The book describes the intelligence stratification of the American society and the resulting emergence of a "Cognitive Elite". The essential conclusions of the book are that more intelligent (higher measured IQ) Americans are selected for college, and end up in fewer professions; American society is becoming cognitively stratified, with the Cognitive Elite crossing paths rarely with those of lower cognitive abilities. In the last half of the twentieth century, more and more Americans have been getting college degrees. College graduates have been funneled into a selective few occupations, especially for the brightest of the bright. The authors assert that more intelligent employees are more proficient employees, so that even among high-IQ professions like law, the highest IQ persons end up at the top.

2.24.1 Bell Curve

The Bell Curve is a controversial best-selling 1994 book by the late Harvard psychologist Richard J. Herrnstein and American Enterprise Institute political scientist Charles Murray. Its central argument is that intelligence is a better predictor of many factors including financial income, job performance, unwanted pregnancy, and crime than parents' socioeconomic status or education level. The bell curve is a Gaussian probability distribution as shown in Figure (2.15). Each country and even state or province has its own curve.

Figure (2.15): Bell curve

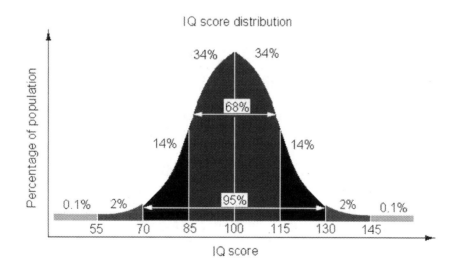

Figure (2.16) shows bell curves for different countries. It depends on the standard level of education, industry, income per capita and other socioeconomical factors.

Figure (2.16): Different bell curves for different countries

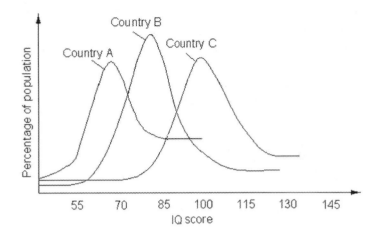

2.25 Intelligence Quotient (IQ)

Wechsler defined intelligence as "The global capacity of a person to act purposefully, to think rationally, and to deal effectively with his/her environment."

The intelligence quotient (IQ) is a score derived from one of several different standardized tests designed to assess intelligence. The term "IQ", from the German Intelligenz-Quotient, was devised by the German psychologist William Stern in 1912 as a proposed method of scoring children's intelligence tests.

IQ scores have been shown to be associated with such factors as:

- Poverty – A low IQ is a strong precursor of poverty, even more so than the socioeconomic conditions in which people grow up. Researchers have long studied ties between poverty and educational achievement, but they're only now beginning to study how poverty affects not only IQ, but brain function and behavior, says University of Pennsylvania neuroscientist Martha Farah
- Schooling -Research has indicated that children who do not attend school or who attend intermittently eventually have poorer scores on IQ tests than those who attend regularly. At the same time, children who move from low-quality schools to high-quality schools are more likely to show improvements in IQ scores. Low IQ raises the likelihood of dropping out of school before completing high school, and decreases the likelihood of attaining a college degree.
- Unemployment, Idleness and Injury - Low IQ is associated with persons who are unemployed, injured often, or idle (removed themselves from the workforce). Some studies underscore the relevance of clinical impairments and neurocognitive deficits to long-term employment and highlight the

need to critically reevaluate the effectiveness of traditional vocational rehabilitation services.

- Family Matters - Low IQ correlates with high rates of divorce, high number of children, lower rates of marriage, and higher rates of illegitimate births. According to a new study, published in the most recent issue of the *Journal Science,* a child's IQ depends significantly on the quality of the family environment that the child experienced during the first two years of life.
- Welfare Dependency - Low IQ increases the chances of chronic welfare dependency. Welfare dependency shifts the bell curve to the left, and accordingly the IQ is lowered.
- Parenting - Low IQ of mothers correlates with low birth weight babies, a child's poor motor skills and social development, and children's behavioral problems from age 4 and up. The behavior of parents and their treatment to their kids has an impact on their cognitive abilities (IQs). Psychologists recognize the need to help parents end the use of corporal punishment and incorporate that objective into their teaching and clinical practice. It is also time for the United States to begin making the advantages of not spanking a public health and child welfare focus, and eventually enact federal no-spanking legislation.
- Crime - Low IQ increases the risk of criminal behavior. The use of IQ as a factor or predictor of violent crime has a long and controversial history. However, study after study has found that many criminals who commit aggressive crimes have lower than average IQs. However, other studies also find a correlation between white-collar crimes, rapist/stalkers and serial murderers and a higher than average IQ.
- Civility and Citizenship - Low IQ people vote least and care least about political and social issues. Success in modern business is typically achieved through a combination of technical skills, timing, character, attitude, and Social IQ. Increasingly, social IQ is one quality which when leveraged can boost success. A workplace where people consistently exhibited their social intelligence and behaved in a way that showed respect for their employers, the organization, co-workers and customers; collaborating with others and sharing your high personal and professional standards for integrity, respect, and civility show high IQ. People with high IQ provide the leaders in your workplace with the tools they need to successfully meet the challenges of constant change, diversity in the workplace, and global communication issues.

2.25.1 Eugenics and IQ

Eugenics is the study and practice of selective breeding applied to humans, with the aim of improving the species. In a historical and broader sense, eugenics can also be a study of "improving human genetic qualities." Advocates of eugenics sought to counter what they regarded as dysgenic (antonym of eugenics) dynamics within the human gene pool. Specifically, in regard to the continuation of congenital disorders and factors impacting overall societal intelligence relating to the heritability of IQ, is not acceptable to them. Advocates of eugenics also justified the necessity for racial and class hierarchy. In order to avoid the danger of committing racial suicide, they proposed two programs; positive and negative eugenic programs to avoid such problems.

The positive program is to promote the production of a better-educated and racially top-quality generation. This program is named by racists who argue that almost all supernatural: blue/blonde/white tone combination is mother natures work of art; all white people are as smart as Isaac Newton; all white people have Isaac Newton brains. The negative program is to stop any increase in the racially unfit through compulsory segregation and sterilization, immigration restriction, and laws to prohibit inter-racial marriage (anti-miscegenation statutes). The racists believe that all white people would now have multiple PhDs in physics, chemistry, mathematics, biology, and neuroscience.

The anti-racists claim that only white people are evil and only white people own slaves and do genocide and all their technology was stolen from coloured people and that science is a white conspiracy made to make non-whites seem inferior, etc. The entire white races get together and have secret meetings to find ways to subjugate these non-Whites (known as the Critical Race Theory).

2.25.2 Heritability of IQ

A *New York Times* op-ed critique describes *The Bell Curve* as "bogus" and "nothing but a racial epithet, because it doesn't consider the combination of genes and environment. Today, it is well accepted that human intelligence is determined by a combination of genes and environment. however what proportion of our intelligence is inherited remains a matter of scholarly debate.

Different studies have measured the heritability of IQ to be anywhere from 40% to 80%. If IQ were only 40% heritable, then only 40% of the variation in human IQ levels would be due to genetic factors, and 60% would be due to environmental factors, such as schooling and diet which heavily influence human intelligence levels. On the other hand, studies that find the heritability of IQ to be closer to 80% suggest that environmental factors have little to do with the broad range of IQs observed in human populations. Such discrepancies as to the role of environment and genetics have led to the argument that ethnic differences in intelligence are due to genetic differences, thus fueling racism debates. These discrepancies have also led to suggestions that early intervention and schooling are a waste of resources.

Herrnstein's and Murray's (bell curve) argument depends on thinking of the 15-point IQ difference (between white and black) as divisible into a genetic chunk and an environmental chunk. This picture suggests the following three alternatives:

Extreme Environmentalism: Blacks are genetically on a par with whites, so the IQ gap is all environmental.

Extreme Geneticism: Blacks are environmentally on a par with whites, so the IQ gap is all genetic.

The Reasonable View: Blacks are worse off both genetically and environmentally: some of the gap is genetic, some environmental.

In general, studies show that covariance between relatives may be due not only to genes, but also to shared environments. Most previous models have assumed different degrees of similarity induced by environments specific to twins, to non-twin siblings (henceforth siblings), and to parents and offspring. The inheritance of intelligence has been investigated for nearly a century. Controversy remains as to how much is inheritable, and the mechanisms of inheritance are still a matter of some debate. Intelligence has also been shown to be associated with such factors as morbidity and mortality and parental social status.

2.25.3 IQ Tests

An intelligence quotient, or IQ, is a score derived from one of several different standardized tests designed to assess intelligence. IQ scores are used in many contexts: as predictors of educational achievement or special needs, by social scientists who study the distribution of IQ scores in populations and the relationships between IQ scores and other variables, and as predictors of job performance and income.

According to Wikipedia, IQ reference charts are tables, suggested by psychologists to divide intelligence ranges into various categories. As reference charts, they are not to be taken as absolute or very precise. The reason for this is the lack of a uniform definition of intelligence and the current inability to wholly quantify it in a scientific manner. To get a "true" IQ score, multiple IQ tests must be taken since professionally administered IQ tests are only roughly 96% accurate. The average of multiple tests, usually at least three, is considered that person's "true" IQ score. However, these tests must be taken within a relatively short period of time, normally under one year for three or more tests.

Generally the IQ scores are estimated and ranked as per the following table; Table (2.3):

Table (2.3): Ranges of IQ

IQ Classifications	IQ Score
Low IQ – Mentally Retarded	70 - 79
Dull Normal IQ	80 - 90
Normal IQ Score	91 – 110
Bright Normal IQ Score	111 – 120
Superior IQ Score	120 – 130
Very superior IQ – Genius IQ Scores	131 and above

Although intelligence quotient (IQ) tests are still widely used in the United States, there has been increasing doubt voiced about their ability to measure the mental capacities that determine success in life.

2.25.4 Types of IQ Tests

There are countless numbers of IQ tests available, with varying amounts of validity. Most commonly used in the United States to measure intelligence in children, the Wechsler Intelligence Scale for Children (WISC I through IV) and Stanford-Binet Intelligence Scales assess an individual's verbal reasoning, abstract and visual reasoning, quantitative reasoning and short-term memory.

Other tests include the Wechsler Adult Intelligence Scale (WAIS-III), Reynolds Intellectual Assessment Scales (RIAS), Universal Nonverbal Intelligence Test (UNIT), Woodcock-Johnson-R, Wechsler Preschool and Primary Scale of Intelligence (WPPSI), Kaufman Brief Intelligence Test (KBIT) and Differential Ability Scales (DAS) for children.

The main ones in use, however, include:

2.25.4.1 WISC-IV (Children)

The Wechsler Intelligence Scale for Children (WISC), developed by David Wechsler, is an intelligence test for children between the ages of 6 and 16 that can be completed without reading or writing. The Wechsler's scales have been widely accepted as a clinical indicator which has both diagnostic and treatment implications.

Wechsler (1958) himself initiated the process of interpreting children's subtest profiles when he advanced the hypothesis that childhood schizophrenia could be diagnosed with the WISC by high scores on Picture Completion and Object Assembly and low scores on Picture Arrangement and Digit Symbol.

Psychologists continued the practice of profile analysis, suggesting that WISC subtest scores could be recategorized to identify children with learning disabilities. Based on factor analytic studies, they suggested that rather than relying on the traditional WISC Verbal and Performance IQs, subtest scores could be redistributed into three "new" composite scores which could identify children with genetic dyslexia. The WISC was revised in 1974 (WISC R) by Wechsler. The Psychological Corporation revised it again in 1991 (WISC III), and again in 2003 (WISC IV). While the WISC R was in use, some researchers discovered the two main factors, the Verbal IQ and the Performance IQ, might be supplemented by a third, labeled Freedom From Distractibility. The WISC III revisions attempted to strengthen this factor, and created a fourth one too. Thus, the WISC III offered a FSIQ as a measure of g, a Verbal and a Performance IQ for those used to the WISC R, along with four new factors. These were the Verbal Comprehension Index, Perceptual Organization Index, Freedom from Distractibility Index, and Processing Speed Index. Of note, while the Psychological Corporation believed the factor structure supported these four factors, Sattler had some doubts.

- Administration of the WISC test
  The Wechsler tests are the most common individually administered IQ tests. They currently include the WISC-IV (age 6-16 years), the WAIS-IV (age 16-89 years), and the WPPSI-III (age 2.5 - 7 years). Shown below in Table (2.4) are the labels and frequency of Wechsler IQ scores. Keep in mind that, due to random factors, IQ scores can vary about 5 points from week to week, and can often change by 10 points or even more over a period of years.

Table (2.4): Labels and frequency of Wechsler IQ scores

IQ	Archaic Description	Description	Score higher than:
10	Idiot	Profound Mental Retardation	Fewer than 1 out of 100,000
25		Severe Mental Retardation	
40	Imbecile	Moderate Mental Retardation	3 out of 100,000
55	Moron	Mild Mental Retardation	13 out of 10,000
70		Borderline	2 out of 100
85	Dull Normal	Low Average	16 out of 100
100		Average	50 out of 100
115		High Average	84 out of 100
125		Superior	95 out of 100

130	Genius	Very Superior/Gifted	98.5 out of 100
145			9,913 out of 10,000

Wechsler IQ tests include two scales: Both the verbal scale has three main subsets, and the performance scale have also three main subsets as shown in Table (2.5).

Table (2.5): Wechsler's scales and subsets

Verbal scales:	
Information (subset), 30%	Measures general knowledge, including questions about geography and literature. Sample question: "What is the capital of Canada?"
Comprehension (subset) , 30%	Measures understanding of social conventions and common sense. Sample question: "What is the thing to do if you find an injured person laying on the sidewalk?"
Information (subset) , 30%	General knowledge questions
Digit Span	Measures concentration, attention, and immediate memory. Requires the repetition of number strings forward and backwards. The digit span subtest requires the child to repeat strings of digits recited by the examiner.
Similarities	Measures verbal abstract reasoning and conceptualization abilities. Some questions: "How are a snake and an alligator alike?"
Vocabulary	Measures receptive and expressive vocabulary. Sample question: "What is the meaning of the word 'malleability'?"
Performance Scales:	
Object Assembly	Consists of jigsaw puzzles. Measures visual-spatial abilities and ability to see how parts make up a whole (this subtest is optional on the revised Weschler tests).
Block Design	Consists of colored blocks which are put together to make designs. One of

	the strongest measures of nonverbal intelligence and reasoning.
Digit Symbol/Coding/Animal House (subset) , 30%	Measures visual-motor speed and short-term visual memory. Symbols are matched with numbers or shapes according to a key.
Picture Arrangement	Measures visual perception, long-term visual memory, and the ability to differentiate essential from inessential details. Requires recognition of the missing part in pictures.
Picture Concepts	Measures non-verbal concept formation and reasoning; a non-verbal counterpart of Similarities.
Matrix Reasoning (subset) , 30%	Measures abstract nonverbal reasoning ability. It consists of a sequence or group of designs, and the individual is required to fill in a missing design from a number of choices.
Arithmetic (subset) , 30%	Measures attention, concentration, and numeric reasoning. Sample question: "John can carry 3 barrels, how many trips he needs to bring 16 barrels?"

In addition to theVerbal Comprehension (VCI), and Perceptual Reasoning (PRI), there are additional components which are Processing Speed (PSI) and Working Memory (WMI).

The WMI's (formerly known as Freedom from Distractibility) subtests are as follows:

- Digit Span - children are orally given sequences of numbers and asked to repeat them, either as heard or in reverse order.
- Letter-Number Sequencing - Children are provided a series of numbers and letters and asked to provide them back to the examiner in a predetermined order.
- Arithmetic (supplemental) - orally administered arithmetic questions that are timed.

The PSI's subtests are as follows:

- Coding - children under 8 mark rows of shapes with different lines according to a code, children over 8 transcribe a digit-symbol code. Time-limited with bonuses for speed.
- Symbol Search - children are given rows of symbols and target symbols, and asked to mark whether or not the target symbols appear in each row.
- Cancellation (supplemental)

2.25.4.2 WAIS-III (the Far Most Commonly Used)

The Wechsler Adult Intelligence Scale (WAIS) intelligence quotient (IQ) tests are the primary clinical instruments used to measure adult and adolescent intelligence.

The WAIS-III, a subsequent revision of the WAIS (the WAIS was initially created as a revision of the Wechsler-Bellevue Intelligence Scale, which was a battery of tests published by Wechsler in 1939) and the WAIS-R (the WAIS-R, a revised form of the WAIS, was released in 1981 and consisted of six verbal and five performance subtests), was released in 1997. It provided scores for Verbal IQ, Performance IQ, and Full Scale IQ, along with four secondary indices (Verbal Comprehension, Working Memory, Perceptual Organization, and Processing Speed).

The sequence of the development of the Wechsler Adult Scale is shown in Figure (2.17).

Figure (2.17): Development of the Wechsler scale

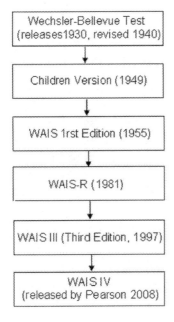

The full scale IQ of WAIS-III consists of Verbal IQ and Performance IQ. A third edition (WMS®–III, it is called Revised Wechsler Memory Scale) was derived to help you examine the important relationship between intellectual functioning and memory. The revised model is also called the 6-Factor Model (Tulsky, Ivnik, Price, & Williams, 2003). Figure (2.18) shows the revised model of WAIS-III.

Figure (2.18): Revised model of WAIS-III

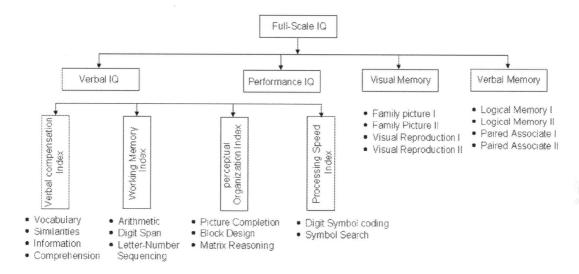

2.25.4.3 Stanford-Binet-IV

The Stanford-Binet intelligence scale (SBS) is a direct descendent of the Binet-Simon scale, the first intelligence scale created in 1905 by psychologist Alfred Binet and Dr. Theophilus Simon. The Stanford-Binet test started with the French psychologist Alfred Binet, whom the French government commissioned with developing a method of identifying intellectually deficient children for their placement in special education programs. This revised edition, released in 1986, was designed with a larger, more diverse, representative sample to minimize the gender and racial inequities that had been criticized in earlier versions of the test.

The test was so popular that Robert Yerkes, the president of the American Psychological Association, decided to use it in developing the *Army Alpha* and the *Army Beta* tests to classify recruits. Thus, a high-scoring recruit might earn an A-grade (high officer material), whereas a low-scoring recruit with an E-grade would be rejected for military service. (Fancher, 1985).

The Stanford-Binet Intelligence Scale (Fourth Edition) has a three-level hierarchical model:
a) The general Intelligence factor (g) at the highest level of interpretation;
(b) Crystallized, fluid, and short-term memory factors at the second level; and
(c) More specific factors such as verbal reasoning, quantitative reasoning, and abstract visual reasoning at the third level.

The Composite score reflects the highest level and is considered to be the best estimate of g in the scale.

The Stanford-Binet scale tests intelligence across four areas: verbal reasoning, quantitative reasoning, abstract/visual reasoning, and short-term memory. The areas are covered by 15 subtests, including vocabulary, comprehension, verbal absurdities, pattern analysis, matrices, paper folding and cutting, copying, quantitative, number series, equation building, memory for sentences, memory for digits, memory for objects, and bead memory.

2.25.4.4 RIAS™ (Reynolds Intellectual Assessment Scales) UNIT

The RIAS is a rapid, valid, comprehensive, and cost-effective assessment of intelligence and memory.

The RIAS is an individually administered test of intelligence appropriate for ages 3 through 94 years, which includes a co-normed, supplemental measure of memory. The RIAS includes:

- A two-subtest Verbal Intelligence Index (VIX), which assesses verbal intelligence by measuring verbal problem solving and verbal reasoning where acquired knowledge and skills are important. The Verbal Memory subtest provides a basic, overall measure of short-term memory skills (e.g., working memory, short-term memory, and learning) and assesses recall in the verbal domain.
- A two-subtest Nonverbal Intelligence Index (NIX) which assesses nonverbal intelligence by measuring reasoning and spatial ability, using novel situations and stimuli that are predominantly nonverbal. The NIX administration of the four intelligence subtests by a trained, experienced examiner requires approximately 20 to 25 minutes. The Nonverbal Memory subtest measures the ability to recall pictorial stimuli in both concrete and abstract dimensions.
- A Composite Intelligence Index (CIX) assesses overall general intelligence, including the ability to reason, solve problems, and learn.

The RIAS provides a thorough assessment of the client's level of intellectual functioning and allows the assessor to evaluate the relationship between the client's memory and cognitive skills. The RIAS scales are ideal for schools and institutions, clinical settings, and for individual practitioners who are looking for ways to control costs and maximize professional time.

The controversy of the RIAS is the argument that the tests are designed by a social class (educated and relatively affluent) whose experiences and vocabulary differ drastically from an inner-city child (for instance) raised in poverty who may be taking the test, thus skewing the outcome.

2.25.4.5 WASI Woodcock-Johnson-R

The WASI includes two verbal subtests (Vocabulary and Similarities) and two nonverbal subtests (Block Design and Matrix Reasoning), each pair of which is used to derive an age-referenced domain IQ score for Verbal or Performance ability, respectively (WASI-verbal, WASI performance). In turn, each of these domain scores contributes to the full-scale IQ score (WASI-full). The WASI Woodcock –Johnson-R (WJR) is used to assess reading fluency and efficiency and includes:

- Oral reading fluency passages
- Listening and reading comprehension
- Aloud reading of passages
- Reading accuracy, acceptability, rate, and expression
- Doing calculations

The above tests can be made differently for each grade. Reading passages at different grades contain different amounts of words, for example:

- 3rd grade – 325 words
- 5th grade – 575 words
- 7th grade – 820 words
- 10th grade – 1010 words

In addition to reading and listening, the Woodcock Johnson-Revised (WJ-R) Calculation has also been used to identify children with MLD (Mathematic Learning Disability), arithmetic, algorithmic computation and arithmetic word problems.

2.25.4.6 WPPSI (Young Children)

The Wechsler Preschool and Primary Scale of Intelligence (WPPSI™) is often used as part of the entrance process for students identified as potentially gifted and talented.

WPPSI-III features shorter, more game-like activities that hold the attention of children as young as 2-1/2 years. The improvement to the *Wechsler* provides more clinically useful information for diagnosis and planning, making WPPSI–III an even more powerful tool.

The WPPSI™-III assesses skills and abilities, rather than grade-level knowledge. The test's propensity toward skills and abilities make for a test-taking experience that is unlike the statewide exams and nationally normed tests (ITBS, Stanford 10, etc.). Thus, preparing for the types of items encountered on the Wechsler Primary is essential; however, the WPPSI is not a test with a defined curriculum for which a child can easily study. You can prepare your children for testing, and beyond the test date, by exposing them to a wide range of thinking activities infused across the curriculum. The idea is to teach children how to think and problem solves, and reflects on their own thinking processes, to improve decision-making skills.

WPPSI-III has undergone substantial revision to increase the scale's age appropriate properties, and to make administration of the scale as user friendly as possible. Both children and examiners benefit from the thoughtful, carefully constructed revisions implemented to build a highly respected, reliable test that completely reflects what customers and other professionals need. The benefits of using WPPSI are listed below:

- Age range has been lowered to 2 years 6 months. Tests of cognitive, motor, behavioral, and neuropsychological function over a period of 30 months can be performed on each band of ages, for example; 6 months, 1 year, 2 years, and 2.5 years.
- There are 5 subtests on the WPPSI-III (age band 2:6-3:11 years). There are 14 subtests on the WPPSI-III (age band 4:0-7:3 years).
- Younger children take fewer subtests that are designed to measure verbal comprehension and perceptual organization abilities
- Older children take a greater number of subtests (seven subtests) designed to measure information, vocabulary, word reasoning, block design, matrix reasoning, picture concepts, and coding. These subtests are assigned to measure perceptual organization, verbal comprehension, and processing speed abilities. In the latest revision, the WPPSI-III was revised with the goal of being more enjoyable for young children and better able to sustain their attention, and to remove any ethnic, gender, regional, or socioeconomic bias.
- Less emphasis on acquired knowledge. However, this test challenges students to remember what has been taught previously in school.
- Instructions to children have been simplified.
- Omission of time bonuses due to the normal delay in motor skill development relative to cognitive skills. Children who were later identified and who had significantly delayed motor skill at all in the first few years of life, could still develop intelligible motor skill at later ages.

The WPPSI-III score summary is divided into five main areas:

1. Full Scale IQ – most reliable and representative of general intellectual functioning. (Information, Vocabulary, Word Reasoning). If the Full Scale IQ score is not valid then either the Verbal or Performance IQ score may be used instead of the Full Scale IQ score.
2. Verbal IQ – acquired knowledge, verbal reasoning and comprehension, and attention to verbal stimuli. Verbal IQ is affected by hyperactivity and conduct problems in children.

3. Performance IQ – fluid reasoning, spatial processing, attentiveness to detail, and visual-motor integration (Block Design, Matrix Reasoning, Picture Concepts). Studies show that hyperactivity in children is also associated with relatively low performance IQ, whereas conduct and emotional problems are not.
4. Processing Speed – ability to quickly and correctly scan, sequence, and discriminate simple visual information.
5. Global Language – expressive and receptive language abilities

2.25.4.7 KBIT (Kaufman Brief Intelligence Test)

KBIT Second Edition gives more information than any other brief intelligence test. Individually administered in just 20 minutes, it assesses both verbal (crystallized) and nonverbal (fluid) intelligence in people from 4 through 90 years of age. Test items are free of cultural and gender bias.

The KBIT-2 is composed of two separate scales:

a. Verbal scale which measures a person's knowledge of word meanings, verbal concept formation, reasoning ability, and a range of general information.
b. Non-Verbal/Matrices which solves novel problems, emphasizes inductive reasoning and visual processing and taps executive functioning to solve problems.

The benefits of KBIT-2 are listed below:

• Features Crystallized (Verbal) Scale
• Measures verbal and nonverbal intelligence quickly
• Is easy to administer and score
• Use for a variety of purposes
• Provides valid and reliable results
• Features high-quality testing materials at a reasonable cost
• Helps to identify students who may benefit from enrichment or gifted programs and
• Identifies high-risk children through large-scale screening who require a more comprehensive evaluation

2.25.4.8 DAS (Differential Ability Scales) for Children

The Differential Ability Scales is an individually administered test battery intending to measure cognitive and achievement levels for children for classification and diagnostic purposes. Psychologists depend on the DAS-II to provide insight into the manner in which a child processes information, giving solutions to fix learning problems. There are four different forms for the test: Preschool, School-Age, Cognitive Battery, and School Achievement.

Approximately 45-60 minutes is required for completion. Its diverse nature makes it possible to profile a child's strengths and weaknesses.

The Differential Ability Scales are an excellent predictor of academic achievement. This instrument is able to address a wide variety of referral questions for a broad age range of children (children, aged 2 years 6 months through 17 years 11 months) in school and clinical settings, as well as in research.

2.25.5 Samples of IQ Tests

2.25.5.1 Verbal IQ Tests

1. to be phlegmatic is to be
    - Apathetic
    - Droopy
    - Industrious
    - Effective
    - I don't know
2. to lampoon is to
    - Mock in a satire
    - catch in an embarrassing situation
    - Remove lighting from a movie set
    - Lie
    - I don't know
3. Imbecile is
    - Moron
    - Superior
    - Affluent
    - Amorphous
    - I don't know
4. If Kids create predicaments to the family, they would
    - Ignore
    - Tolerate
    - Punish
    - Remit
    - I don't know
5. When food is copious, prices would be
    - Exorbitant
    - Enigma
    - Vying
    - Cost-effective
    - I don't know

Choose the word or phrase that is most similar to the one given.*
6. Ribald
    - Coarsely funny

- o Having a shared head
- o Old tasting
- o Agitate
- o I don't know
7. Fabulist
    - o Charmer
    - o Funny person
    - o Liar
    - o Impartial
    - o I don't know
8. Antic
    - o Thief like
    - o Cat like
    - o Clown like
    - o Pig like
    - o I don't know

Choose the pair that best expresses a relationship similar to that reflected in the question.*

16. Vertigo: Dizzy
    - o Temperature: Virus
    - o Influenza: Fervish
    - o Fear: Heights
    - o Illness Germs
    - o I don't know
17. Indigenous: native
    - o New: Used
    - o Alien: Foreign
    - o Aged: Wrinkled
    - o Immigrate: Exodus
    - o I don't know
19. Paint: Canvas
    - o Brush: Bucket
    - o Peanut butter: Bread
    - o Clock: Time
    - o Art: Life
    - o I don't know

Choose the word(s) that best fit in the sentence.*

20. The city's feeble art scene involved a couple of _____ museums, poorly lit, badly designed and stuffed with the amateurish paintings of dabblers and _____.*
    - o grand…demagogues
    - o somniferous…dilettantes
    - o resonant…parsimonious fools

o   vaulted…gastronomes
o   I don't know

23. Since Tom had been working with people intensively all week, he craved a quiet weekend of _____to give himself a chance to read, rest, and relax.*

o   alone
o   solitude
o   congregation
o   enterprise
o   I don't know

The above assessment is meant for those whose first language is English

* From http://psychologytoday.tests.psychtests.com/bin/transfer

2.25.5.2 Non-Verbal IQ Test

1)

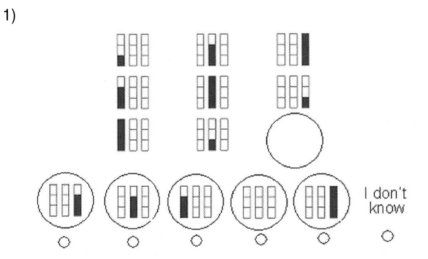

2)

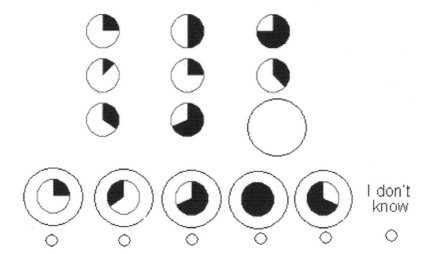

I don't
know

3)

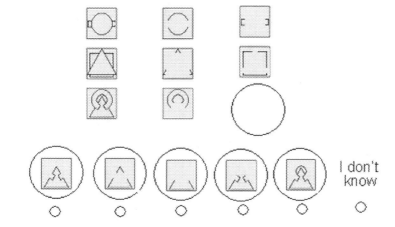

I don't
know

4)

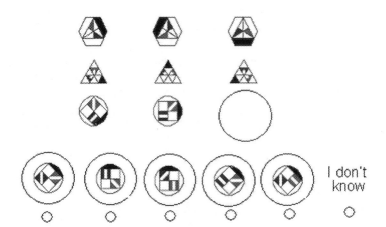

I don't know

5)

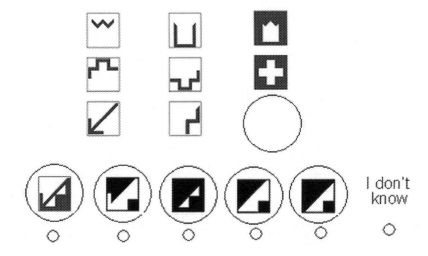

I don't know

6)

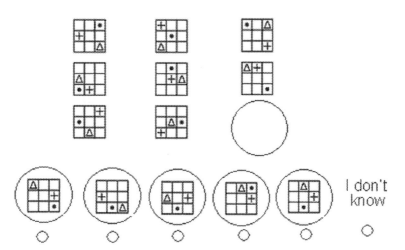

7)

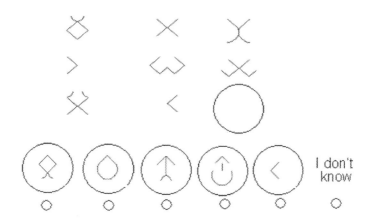

2.25.5.3 Performance Test

1) Two of the following numbers add up to eighteen.   5 - 6 - 2 - 13 - 7 - 15

   O true   O False

2) Which one of the five is least like the other four?   Whale Lizard Snake Tiger Frog.

3) The circles below represent the orbits of three planets. Which region represents all three orbits: Planet A, Planet B, and Planet C.?

   O   a
   O   b

O   c
O   d
O   e
O   f
O   g

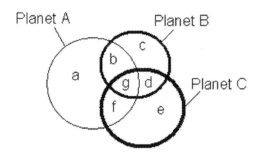

Planet A   Planet B

Planet C

4) If you rearrange the letters "PSNIA", you would have the name of a/an:

O   Country
O   City
O   Ocean
O   State
O   Continent

5) Jim travels five miles north, then three miles east, then four miles south and finally three miles west.   How many miles is Jim from his starting place?

O   1 mile
O   2 miles
O   3 miles
O   4 miles

6) Which one of these four shapes is least like the other three?

O   a
O   b
O   c
O   d

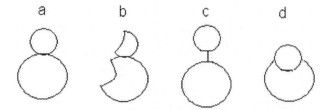

a   b   c   d

7) What is the minimum number of toothpicks necessary to spell the word "LAWN"? (You are not allowed to break or bend any toothpicks.)

- O 10
- O 11
- O 12
- O 13
- O 14

8) Assume that the area of block a and the area of block b form the area of block c exactly. Also assume that the area of the block a in square centimeters is three times the area of the block b. If the area of the block a is 120 square centimeters, what is the area of the block c?

9) Jim's grandfather's son is Fred's dad.
How is Fred related to Jim? (Assume nobody has ever remarried.)

- O Fred is Jim's uncle
- O Jim is Fred's uncle
- O Jim is Fred's father
- O They are first cousins

10) Antagonism is to agnosticism as 3653ag6pt4 is to
Considering the letter c as m, write the answer in below.

----

11) At a store, they cut the price 25% for a particular item. By what percent must the item be increased if you wanted to sell it at the original price?

- O 33.33%
- O 66.66%
- O 40%
- O 45%
- O 47%

12) Which one of the numbers does not belong in the following series: 1 3 7 15 32 63 127

- O 7
- O 15
- O 32
- O 63
- O 127

13) Jim, twelve years old, is twice as old as his brother. How old will Jim be when he is one and a half times as old as his brother?

O 16
O 17
O 18
O 21
O 24

14) If the area of block a is 16 centimeters, what is the approximate area of blocks b, c and d?

15) Jim received $0.66 change from a purchase in the drugstore. If he received six coins, and three of the coins are the same denomination, how many quarters did he receive?

O two
O three
O one
O none
O The question can't be answered

2.26 Decline or Increase in IQ Scores

IQ can be affected by any number of things (parenting, poverty, environmental factors, and school attendance, among others), but is statistically not likely to increase over one's lifetime. One's IQ is, for the most part, unchanging after about the age of 7, [http://www.brainy-child.com/expert/iq-score.shtml].

Some psychologists argue that IQ can be changed, though not drastically in a short period. Especially during infancy and early childhood, there is a possibility of a change in IQ scores frequently. However, IQ scores begin to stabilize in middle childhood. Furthermore, by the age of approximately 7 years, childhood IQ scores are found to be rather good predictors of adult IQ scores.

As we age, after the age of 30, some of our mental ability and performance may be affected and this would be reflected in our IQ scores. With this, good IQ test scores take into account changes in mental ability over age by modifying the IQ score, which in turn, shows little decline in IQ scores (assuming other factors remain the same).

2.26.1 Factors Affecting the IQ Scores:

- Parenting: Recent studies suggest that kids born to older dads face a higher chance of developing bipolar disorder. Apparently children with older dads may have lower IQs, too. According to Dr. John McGrath from the Queensland Brain Institute at the University of Queensland in Brisbane, Australia, and the study's lead researcher, more mutations occur in men's sperm as they age, and those mutations could impact brain development.

  Recent studies also show that children's genetic makeup affects their own behavioral characteristics, and also influences the way they are treated by their parents. Children's genetic predispositions and their parents' childrearing regimes are seen to be closely interwoven, and the ways in which they function jointly can affect the IQ of their children.

- Poverty and life style: Poverty and life style (such as poor nutrition), low maternal education, unemployment, inadequate living conditions, limited educational opportunities, and poor health can raise the risk of poor intellectual development. Social Stratification, social classes and meritocracy can also affect the IQ level.

- Environmental factors: For the last century, scholars have been preoccupied with identifying the exact factors that influence one's IQ. The numerous studies on the subject have led most scientists nowadays to the belief that one's IQ is determined by a variety of both genetic and environmental factors, although there is contention about the exact weight of each. The majority of studies on intelligence have shown that environmental factors account for about 25% of the differences in people's IQ scores.

- Evidence showed that Identical twins reared apart have IQ's that are less similar than identical twins reared in the same environment (McGue & others, 1993). Also, school attendance has an impact on IQ scores (Ceci, 2001). Children who are breastfed during the first three to five months of life score higher on IQ tests at age 6 than same-age children who were not breastfed (Reinberg, 2008). Scientists have discovered many factors during a woman's pregnancy that could affect a child's cognitive development. Such factors include the mother's health, her nutrition and smoking and drinking habits during pregnancy, her age and the number of previous pregnancies, the interval since her last pregnancy, the blood type and Rh incompatibility of mother and fetus, her history of X-ray exposure, and her red blood cell count, to list a few (Jensen; 169).

- School attendance: Studies have shown that the length of staying at school affects the IQ and cognitive ability. Studies also show that the IQ score declines over the holiday months. This proves that each day away from school especially for children whose vacations are least academically orientated, decline is seen, however, rather marginal and possibly temporary.

- Different tests: Different tests may indicate different scores. However, it is expected that standardized and reliable tests should not show much difference. Different tests also may test different sets of cognitive abilities. IQ tests may also differ from one country to another.

2.27 General Ability Index (GAI)

The general ability index (GAI) is a composite score that is based on number of verbal comprehension and number of perceptual reasoning subtests. The number can be 3 or 4. It does not have to include the working memory or processing speed subtests included in the full scale IQ (FSIQ). The original Wechsler Intelligence Scale for Children (WISC; Wechsler, 1949), the Wechsler Intelligence Scale for Children—Revised (WISC–R; Wechsler, 1974), and the WISC–III included an FSIQ as well as a verbal IQ (VIQ) and a performance IQ (PIQ).

2.28 Gifted Children and Individuals

IQ test scores are now commonly used for identifying gifted and talented children. A child who scores above 130 on the intelligence test is considered to be gifted. The intelligence test measures rational intelligence, which is divided into 3 types:

- Verbal (linguistic) intelligence
- Logical (mathematical) intelligence
- Visual (spatial) intelligence

2.28.1 Characteristics and Behaviour of the Gifted

- Many gifted children learn to read early, with better comprehension of the nuances of language. They show skill in drama/art/music/language
- Gifted children often read widely, quickly, and intensely and have large vocabularies. They demonstrate strong expressive skills and elaborate on ideas.
- They can work independently at an earlier age and can concentrate for longer periods.
- Gifted children can have hidden learning disabilities that go undiscovered because they can easily compensate for them in the early years.
- Gifted students are capable of producing high level products in specific areas of learning at the level of a competent adult

- Gifted children commonly learn basic skills better, more quickly, and with less practice.
- They often have seemingly boundless energy and can concentrate for longer periods.
- They exhibit an intrinsic motivation to learn, find out, or explore and are often very persistent. "I'd rather do it myself" is a common attitude.
- Gifted children like to learn new things, are willing to examine the unusual, and are highly inquisitive.
- They usually respond and relate well to parents, teachers, and other adults. They may prefer the company of older children and adults to that of their peers.
- They may read a great deal on their own, preferring books and magazines written for children older than they are.
- They may show keen powers of observation and a sense of the significant; they have an eye for important details.
- They are often skeptical, critical, and evaluative. They are quick to spot inconsistencies.
- They often have a large storehouse of information about a variety of topics, which they can recall quickly.
- They are elaborate thinkers, producing new steps, ideas, responses, or other embellishments to a basic idea, situation, or problems.
- Giftedness can be observed in the first three years by rapid progression through the developmental milestones.
- The ideal age for testing is between 5 and 8 ½ years. By the age of 9, highly gifted children may hit the ceiling of the tests, and gifted girls may be socialized to hide their abilities. Unless they are absolutely certain they are right, gifted girls are often unwilling to guess, which lowers their IQ scores.
- When one child in the family is identified as gifted, the chances are great that all members of the family are gifted.
- Second children are recognized as gifted much less frequently than first-borns or only children. Even the first-born identical twin has a greater chance of being accepted in a gifted program than the second-born!
- IQ testing in childhood clearly demonstrates the equality of intelligence between males and females. Until the IQ test was developed, most of society believed in the "natural superiority of males." Even now, the fact that most of the eminent are men leads some to believe that males are innately more intelligent than females. Table (2.6) shows the IQ for male and female in different periods.

Table (2.6): The difference of IQ between males and females in different periods*

	Males above 160 IQ	Females above 160 IQ	Total

1979 – 1989	94	89	183
1990 – 2009	507	298	805
1979 – 2009	601	387	988

* http://www.gifteddevelopment.com/What_is_Gifted/learned.htm

- Gifted individuals also experience the world differently, resulting in certain social and emotional issues. The work of Kazimierz Dabrowski suggests that gifted children have greater psychomotor, sensual, imaginative, intellectual, and emotional "overexcitabilities".
- Gifted children enjoy multiple intelligences such as linguistic, logic-mathematical, musical, spatial, kinesthetic, interpersonal, intrapersonal, naturalistic and existential.
- Giftedness is frequently not evenly distributed throughout all intellectual spheres; an individual may excel in solving logic problems and yet be a poor speller; another gifted individual may be able to read and write at a far above average level and yet have trouble with mathematics.

2.28.2 Environment and Genetics Contribution to Giftedness

Intellectual giftedness is a very complicated human phenomenon, influenced by genetic and environmental factors, but with a distinctive extent during personal growth. Multiple factors affect giftedness. These factors include emotional characteristics, social knowledge and relationships and environment.

Some theorists have argued for a genetical explanation of intelligence, one which places heavy emphasis on the genetic contribution and less weight on environmental contributors (Terman, 1975; Scarr-Salapatek, 1976). Other theorists (environmentalists) suggested that children can be gifted if they are given the proper stimulation and early teaching. Most of us probably believe in a midline position between these two extremes and consider that the identification and understanding of giftedness requires inspection of both child characteristics and environmental factors. The child's behavioral style and interaction with the environment's response has been suggested as likely to account for a portion of the variance in growing up gifted (Lewis & Michalson, 1985; Lewis & Coates, 1979).

Recent studies used the Gardiner's model to differentiate between responsibilities of genetical and environmental skills, Table (2.7).

Table (2.7): Genetical and environmental responsibilities on intelligence

Skill	Meaning	Genetical/Environmental
Interpersonal	Interacts with others, empathizes, socializes, and mediates conflicts	Genetical and Environmental
Intrapersonal	Focuses inwards on dreams and feeling, follows instincts	Genetical and Environmental
Naturalists	Understands and appreciates environment, believes in evolution	Environmental
Logics and mathematics	Enjoys logics, mathematics, reasoning and problem solving	Genetical (Prefrontal Cortex)
Linguistics	Enjoys manipulating languages, speaking and writing, poems	Genetical (Left Frontal Lobe)
Spatial	Creates mental images, enjoys solving puzzles, and visualization tasks	Genetical (Right Parietal Lobe) and environmental
Bodily/Kinesthetic	Enjoys athletic and physical activities	Genetical (Motor Cortex and cerebellum) and environmental
Musical	Enjoys musing and picking up sound and rhythem	Genetical (Temporal lobe) and environmental

2.28.3 Intelligence of Prominent Figures

1. Einstein was four years old before he could speak and seven before he could read.
2. Isaac Newton did poorly in grade school.
3. Louis Pasteur was rated as mediocre in chemistry when he attended the Royal College
4. Bill Gates didn't seem like a shoe-in for success after dropping out of Harvard and starting a failed first business with Microsoft co-founder Paul Allen called Traf-O-Data. While this early idea didn't work, Gates' later work did, creating the global empire that is Microsoft.
5. A newspaper editor fired Walt Disney because he had "No good ideas"
6. F.W.Woolworth got a job in a dry goods store when he was 21. But his employers would not let him wait on a customer because he "Didn't have enough sense."
7. Abraham Lincoln entered The Black Hawk War as a captain and came out a private.
8. Winston Churchill failed the sixth grade.
9. Robert Sternberg (big name in psychology) received a C in his first college introductory psychology class with his teacher telling him that, "there was already a famous Sternberg in psychology and it was obvious

there would not be another." Sternberg showed him, however, graduating from Stanford with exceptional distinction in psychology, summa cum laude, and Phi Beta Kappa and eventually becoming the President of the American Psychological Association

10. Caruso's music teacher told him "You can't sing, you have no voice at all."
11. Leo Tolstoy flunked out of college.
12. Verner Von Braun flunked 9th grade algebra.
13. Thomas Edison: In his early years, teachers told Edison he was "too stupid to learn anything." Work was no better, as he was fired from his first two jobs for not being productive enough. Even as an inventor, Edison made 1,000 unsuccessful attempts at inventing the light bulb. Of course, all those unsuccessful attempts finally resulted in the design that worked.
14. Admiral Richard E. Byrd had been retired from the navy, as, "Unfit for service" Until he flew over both poles.
15. Dick Cheney: This recent Vice President and businessman made his way to the White House but managed to flunk out of Yale University, not once, but twice. Former President George W. Bush joked with Cheney about this fact, stating, "So now we know –if you graduate from Yale, you become president. If you drop out, you get to be vice president." George W. Bush was graduated from Yale University.
16. Fred Waring was once rejected from high school chorus.
17. Sidney Poitier: After his first audition, Poitier was told by the casting director, "Why don't you stop wasting people's time and go out and become a dishwasher or something?" Poitier vowed to show him that he could make it, going on to win an Oscar and become one of the most well-regarded actors in the business.

2.29 Different Forms of Giftedness (Multiple Talented Savant)

Gifted children can be simply identified by simple observation of the child's behaviour by an educational professional, a parent or friend. Some subjective elements are certainly involved particularly in comparisons with other children of the same age.

The following characteristic traits are listed by a broad category of giftedness. It is not expected that any gifted child will share all the traits:

- General intellectual ability (shows avid interest in science or literature, tends to dominate peers or situations, shows superior judgment in evaluating things, asks many questions of a provocative nature, makes good grades in most subjects, etc).
- Creative thinking and production (provides multiple solutions or responses to problems, is fluent in producing and elaborating on ideas, acts spontaneously, intuitively, asks provocative questions, challenges parents, teachers, written and other authorities, is bored with memorisation and recitation, etc).

- Specific academic attitude (learns rapidly, easily and with less repetition in a few specific areas or all areas, is able to show broad perspective on one or more subjects, likes to study some subjects more than others, etc).
- Psychomotor ability (is coordinated, balanced and confident in physical activities, is rhythmic, athletic and energetic, demonstrates endurance, stamina and persistence in physical activities, etc).
- Leadership (can articulate ideas clearly, can stimulate and arouse others, interacts with others easily showing social skills, can establish the mood of a group, is often asked for ideas and suggestions, etc).

2.29.1 Autism Spectrum Disorder (ASD)

Terms used to refer to various Autism Spectrum Disorders (ASDs) can be very confusing and can have terms such as:

- Autism, classic autism, high functioning autism
- Autistic Disorder (severe)
- Asperger Syndrome or Asperger's Disorder (milder form)
- Pervasive Developmental Disorder (PDD), and/or Autism Spectrum Disorder (ASD)

Autism disorder changes the way the brain processes information and can affect all aspects of a person's development. Classic autism usually appears during the first three years of life. Autism is four times more common in boys than girls. Autism Spectrum Disorder is a neurological disorder resulting in developmental disability. This affects communication, social understanding, behaviour, activities and interests. Other aspects of autism spectrum disorder, such as atypical eating, are also common but are not essential for diagnosis; they can affect the individual or the family. Making and maintaining friendships often proves to be difficult for children with autism. For them, the quality of friendships, not the number of friends, predicts how lonely they are, despite the common belief that they prefer to be alone. Being on the autism spectrum does not keep children from understanding race and gender stereotypes in a society; like normal children they learn stereotypes by observing their parents' actions, Hirschfeld L, Bartmess E, White S, Frith U. Can autistic children predict behavior by social stereotypes? *Curr Biol.* 2007; 17(12):R451–2.

Some individuals with ASD show unusual abilities, ranging from splinter skills such as the memorization of trivia to the extraordinarily rare talents of prodigious autistic savants, Treffert DA, Wisconsin Medical Society. Savant Syndrome: an extraordinary condition – a synopsis: past, present and future; 2006.

There is no single best treatment package for all children with ASD. Decisions about the best treatment, or combination of treatments, should be made by the parents with the assistance of a trusted expert diagnostic team.

The exact causes of ASDs aren't fully understood at present. Some research indicates that the genes we inherit from our parents play an important role. Others partially relate the disease to the environment. Many different genes have been implicated in ASDs, suggesting that ASDs have a variety of causes.

Children with Asperger syndrome usually have above-average intelligence. Some children are skilled in fields requiring logic, memory and creativity, such as maths, computer science and music. They show an intelligence of an IQ higher than 130 in some topic such as gifted drawers, gifted musicians, gifted artistics, gifted calendrical calculators and globally gifted children.

2.29.2 Weak Central Coherence (WCC)

The weak central coherence hypothesis of Frith is one of the most prominent theories concerning the abnormal performance of individuals with autism on tasks that involve local and global processing. Individuals with autism often outperform matched nonautistic individuals on tasks in which success depends upon the processing of local features, and underperform on tasks that require global processing. For example, children with WCC can see the leaves of the tree, but they don't see the wood.

The weak central coherence theory attempts to explain how some people diagnosed with autism can show remarkable ability in subjects like math, art, drawing, calendar calculation and engineering, yet have trouble with language skills and tend to live in an isolated social world.

2.29.3 Different Forms of Giftedness

2.29.3.1 Gifted Drawers

Gifted drawers don't depict a generic picture/object, but include a rich amount of detail. They draw recognizable, complex images quickly and with ease. They don't labor over and erase their lines and curves. They can start a drawing from just about any part of the object drawn. Picasso could draw a dog by starting with the tail or the ear (not the usual starting point), with no decrement in speed or confidence. Gifted drawers may well be using a figural, perceptual strategy, rather than a conceptual one. That is, they just see the shapes of things, including the distortion of shapes as they recede into depth and diminish in size or become foreshortened. They don't need to rely on a learned system or perspective drawing. Typically, they don't begin to draw in perspective until late childhood or even adolescence.

Drawing is by definition a perceptual-motor function which relates to drawing aptitude. Selfe(1983) suggested that drawers followed a similar pathway of perceptual motor function to that found in typical development but at a much faster and more efficient rate (Golomb, 2004). The exceptional representational drawing ability in cases of autism may be because advanced skill in visual/spatial skills have become particularly enhanced in these children because their right side of the brain (which has a major responsibility in visual/spatial processing) may compensate for deficits in their left side of the brain (which has a major responsibility for language and verbal tasks). Alternatively, it may be more appropriate to understand their exceptional representational drawing ability as an artistic skill they possess rather than a result of a cognitive deficit associated with their autism (Golomb, 2004).

The changes in visual/spatial representation that occur during childhood have created interest among psychologists, therapists, teachers and parents. The interpretation of these changes has been significant in estimating the stage of development, I.Q., personality and emotional disturbance.

2.29.3.2 Verbally Gifted Children

The term verbally gifted children is used to refer to children who have strong language skills. Verbally gifted kids become competent in language before their age mates do. They are definitely well advanced for their ages in terms of vocabulary and ability to construct full sentences. They also perform better on verbal and general information tests and tests of English (or their own languages) expression than mathematically gifted children do.

Verbally gifted children show unique performances in:

- Writing of fictions or nonfiction, including song lyrics
- Enjoying literary discussions
- Enjoying reading, no matter how specialized the book is
- Showing improvisational abilities in memorizing, recitation, and vocabulary recall
- Enjoying reading journals, blogs, papers and newspapers
- Answering and solving all assignment on time with no challenges

Scientist at Johns Hopkins University in Baltimore say that children who possess extremely high levels of mathematical or verbal ability tend, far more often than children of normal ability, to be left-handed, nearsighted and suffering from asthma or other allergies.

2.29.3.3 Mathematically Gifted Children

One of the cognitive neuroscience literatures proposed the brain characteristics of math giftedness is the enhanced development of the right cerebral hemisphere (RH) and an unusual reliance on its specialized visuospatial processing capacities (Geschwind & Galaburda, 1984; O'Boyle, Benbow, & Alexander, 1995). Another is a special form of brain bilateralism (O'Boyle et al., 2005), involving heightened connectivity and integrative exchange of information between the left and right cerebral hemispheres (O'Boyle & Hellige, 1989; Singh & O'Boyle, 2004).

The term mathematically gifted is used to refer to children who have strong mathematical abilities or those who engage in qualitatively different mathematical thinking. Mathematically gifted children focus on and have interest in mathematical things such as numeric and/or geometric. The mathematically gifted kids perform better on spatial, nonverbal reasoning, speed, memory, and mechanical comprehension tests than verbally gifted children do. Children gifted and talented in mathematics tend to view the world through a mathematical lens.

Mathematically gifted children are characterized in the following ways:

- They are high achievers and are fast and accurate workers
- they prefer to work in homogeneous groups
- Gifted students need little teacher support
- They learn differently from other age group peers. They require curriculum to be differentiated to meet their specific needs
- They are able to see relationships among topics, concepts, and ideas without the intervention of formal instruction specifically geared to that particular content

2.29.3.4 Calendrical Calculators

Calendrical calculators posses the autistic savants skill of being able to calculate the day for any given date. While they may not be able to do relatively simple arithmetic they have this specific ability which, if achieved mathematically, requires complex operations, as O'Connor (1984) reports. Calendrical calculation is the talent of naming days of the week corresponding to dates in the past, present and future. It is puzzling why anyone should develop it, as it has no obvious purpose or value. Remarkably, most reported cases are individuals with below average intelligence (or in some cases above average intelligence). Among these are several whose exceptional level of skill marks them as prodigious savants (Treffert, 1989). They can accurately answer date-questions over a large range of years, i.e., 50 years or more, taking less than 10 s for each question. The existence of calendrical calculation skills and other savant talents such as art and music have been taken to show the independence of various forms of intelligence (Gardner, 1983). Calendrical calculators could be considered under the category of globally gifted children (see the section of globally gifted children).

The scarcity of the talent may have masked relations between calendrical calculation and general intelligence. To detect such relations requires a sample of calendrical calculators. Several aspects of calendrical calculation might depend on intelligence. In common with those possessing other complex cognitive skills, date-calculators may vary in range, accuracy and latency. Previous case reports suggest that savant calculators differ considerably in range. George (Horwitz, Deming, & Winter, 1969; Horwitz, Kestenbaum, Person,

& Jarvik, 1965) had a range of more than 40,000 years while B, studied by Hill (1975), had a range of 32 years. Hermelin and O'Connor (1986) found calculators with higher Wechsler Intelligence Test IQs were more accurate and could answer questions about future dates. While Young and Nettelbeck (1994) did not confirm these results, they only had data for three calculators whose Wechsler IQs were similar, i.e., ranging from 65 to 76, [http://www.accessmylibrary.com/article-1G1-61892312/calendrical-calculation-and-intelligence.html]

To conclude:

- There is no accepted summing up of a gifted child. As with all kids, each one is unique in his/her giftedness/talent.
- Some children exhibit "domain-specific giftedness" where they may excel in one or more particular areas.
- In fact, gifted kids can have below average abilities (or learning disabilities) in some areas while being gifted in other areas.

2.29.3.5 Musical Giftedness

Musically gifted children have a keen ear for music, can play demonstrated or familiar pieces beautifully, have an extraordinary sense of rhythm, and are well synchronized with their peers when playing their instruments.

The Association for Supervision and Curriculum Development and the other Heritage Songwriters Associations report that children who are musically gifted show early developmental signs of musical precocity, which may include noticing off-key music, remembering melodies, singing in tune, fondness for playing instruments in preschool, rhythmic ways of moving and speaking, humming to themselves, tapping rhythmically while working, and sensitivity to environmental sounds (waterfalls, rain on the roof, thunder, etc.).

Musically gifted children:

- match pitches accurately
- are able to duplicate complex rhythms correctly
- demonstrate unusual ability on instruments including voice
- have a high degree of aural memory/musical memory
- display compulsive musical pursuit
- can be aware and discriminate perceptual attentiveness of sounds, rhythmic sense and sense of pitch
- can be aware of the interaction between the listener and the performer
- Once they exceed 90 in the IQ, a high IQ is irrelevant in the fields of music and art.

2.29.3.6 Globally Gifted Children

The term globally gifted is used to refer to children who are evenly gifted or gifted in all or most areas, usually in both math and verbal.

Brains of the globally gifted children are atypical. Their heads tend to be larger, their reflexes in all disciplines are faster, and their brains show atypical brain scan patterns. Their brain structure, brain size, brain speed, brain efficiency, bilateral representation of language and language-related problems are different from normal children. They are left-handed and exposed to immune system disorders.

As with a disability, global giftedness can lead to unhappiness and social isolation. With adult minds in children's bodies, extremely gifted children tend to be maltreated by other children. They tend to find little unity and harmony with their age peers, relating to older children or adults.

2.29.3.7 Savants

Current research theories show that intelligence is non-local and not bound to the brain. Sometimes nature offers perceptive insight into a particular subject by presenting a baffling enigma and contradictory example. Intelligence's contradictory enigma is the idiot-savant.

A savant is a learned person, well versed in literature or science, often with an exceptional skill in a specialized field of learning. The term is also commonly used as an acronym of an autistic savant, formerly "idiot savant". The word idiot usually refers to a simpleton, in contrast to the word "savant" in French that means "learned one." Idiot savants are a subgroup of a class of people called idiots with an IQ of about 25.

An autistic savant (it used to be named "idiot savant") is a person with extraordinary mental abilities, often in numerical calculation, but sometimes in art or music. These skills are often, yet not always, associated with autism or mental retardation. Idiot savants are a group of humans that are incapable of learning, writing or reading, yet they have unlimited access to specific, accurate knowledge in the fields of mathematics, music, and other precise areas. Now the irony of an idiot-savant is that this group of individuals does not acquire knowledge by learning as the average human does.

Let's take a look at one savant with superhuman mental skills.

Kim Peek, the Real Rain Man (Wikipedia)

Kim Peek was born in Salt Lake City, Utah with macrocephaly damage to the cerebellim, and agenisis of the corpus callosum (a condition in which the bundle of nerves that connects the two hemispheres of the brain is missing) In Peek's case, secondary connectors such as the anterior commissure were also missing. There is speculation that his neurons made unusual connections due to the absence of a corpus callosum, which resulted in an increased memory capacity.

His childhood doctor told Kim's father to put him in an institution and forget about the boy. Kim's severe developmental disabilities, according to the doctor, would not let him walk let alone learn. Kim's father disregarded the doctor's advice.

Till this day, Kim struggles with ordinary motor skills and has difficulty walking. He is severely disabled, cannot button his shirt and tests well below average on a general IQ test.

But what Kim can do is astounding: he has read some 12,000 books and remembers everything about them. "Kimputer," as he is lovingly known to many, reads two pages at once - his left eye reads the left page, and his right eye reads the right page. It takes him about 3 seconds to read through two pages - and he remembers everything on them. Kim can recall facts and trivia from 15 subject areas from history to geography to sports. Tell him a date, and Kim can tell you what day of the week it is. He also remembers every music piece he has ever heard.

Since the movie *Rain Man* came out, Kim and his father have been traveling across the country for appearances. The interaction turns out to be beneficial for him, as he becomes less shy and more confident.

In 2004, NASA scientists examined Kim with a series of tests including computerized tomography and magnetic resonance imaging. The intent was to create a three-dimensional view of his brain structure and to compare the images to MRI scans done in 1988. These were the first tentative approaches in using non-invasive technology to further investigate Kim's savant abilities. A 2008 study concluded that Peek probably had FG syndrome, a rare genetic syndrome linked to the X chromosome which causes physical anomalies such as hyptonia (low muscle tone) and macrocephaly (abnormally large head). It was discovered that Peek had no corpus callosum; the part of the brain that links the two hemispheres. The corpus callosum is the structure deep in the brain that connects the right and left hemispheres of the cerebrum, coordinating the functions of the two halves.

2.30 Emotional Intelligence

Emotional intelligence (EQ, EI, or EIQ) is the innate potential to feel, use, communicate, recognize, remember, describe, identify, learn from, manage, understand and explain emotions. - S.Hein, 2007

Emotional intelligence is based on an innate potential and the effect of the environment over a person's life. Environmental effect could have a large effect on the emotional intelligence.

Many definitions of Emotional Intelligence have been suggested:

1- Daniel Goleman suggested in his best selling book (1995) *Emotional Intelligence* that emotional intelligence (EQ) refers to the ability to recognize and regulate emotions in ourselves and others. In his book, emotional intelligence addressed our response to every day social interaction and challenges.

2- Peter Salovey and John Mayer defined in their published writing the term "emotional intelligence" as :

> *A form of intelligence that involves the ability to monitor one's own and others' feelings and emotions, to discriminate among them and to use this information to guide one's thinking and actions (Salovey & Mayer, 1990).*

Later, these authors revised their definition of emotional intelligence, the current characterization now being the most widely accepted. Emotional intelligence is thus defined as:

> *The ability to perceive emotion, integrate emotion to facilitate thought, understand emotions, and to regulate emotions to promote personal growth (Mayer & Salovey, 1997).*

3- Reuven Bar-On is another prominent researcher (the originator of the term "emotion quotient") who defined emotional intelligence as being concerned with understanding oneself and others, relating to people, and adapting to and coping with the immediate surroundings to be more successful in dealing with environmental demands (Bar-On, 1997).

Emotional intelligence describes the ability, capacity, skill or, in the case of the trait EQ model, a self-perceived grand ability to identify, assess, plan, manage and control the emotions of one's self, of others, and of groups. It is some times termed "social intelligence". Emotional Intelligence is increasingly relevant to organizational development and personal development, because the EQ principles provide a new way to understand and assess people's behaviours, management styles, attitudes, interpersonal skills, and potential. It is an important factor in human resources planning, job profiling, recruitment interviewing and selection, management development, customer relations and customer service, and more.

The EQ concept argues that IQ, or conventional intelligence, is too narrow; that there are wider areas of Emotional Intelligence that dictate and enable how successful we are. Success requires more than IQ (Intelligence Quotient), which has tended to be the traditional measure of intelligence, ignoring essential behavioural and character elements. We've all met people who are academically brilliant and yet are socially and inter-personally inept. And we know that despite possessing a high IQ rating, success does not automatically follow.

With emotional intelligence, one has to be honest and perceive things according to what he/she really does, feels, or thinks, rather than what he/she thinks is considered right. Nobody can judge you, except yourself. Every person has his/her own perception.

Each child is born with a specific and unique potential for these components of emotional intelligence:

- Emotional memory
- Emotional processing ability and problem solving ability
- Emotional sensitivity
- Emotional learning ability

The way a child is raised dramatically affects what happens to his/her potential in each of the above components. For example, a child might be born with a very high potential for mathematics -- he or she might be a potential Isaac Newton -- but if that child's potential is never recognized, nurtured, and encouraged, and if the child is never given the chance to develop their mathematical potential, they will never become a talented mathematician later in life. The world will then miss out on this person's special gift to humanity. Also, a child being raised in an emotionally abusive environment can be expected to use their emotional potential in detrimental ways later in life.

Goleman identified the five 'domains' of EQ as:

1. Knowing your emotions

2. Managing your emotions

3. Motivating yourself

4. Recognizing and understanding other people's emotions

5. Managing relationships, i.e., managing the emotion of others

Goleman summarized the emotional intelligence in two aspects*:

1. Personal competence

   • Self-awareness

     a. Emotional awareness - recognizing one's emotions and their effects.

     b. Accurate self-assessment - knowing one's strengths and limits.

     c. Self-confidence - sureness about one's self-worth and capabilities.

   • Self-regulation

     a. Self-control - managing disruptive emotions and impulses.

     b. Trustworthiness - maintaining standards of honesty and integrity.

     c. Conscientiousness -taking responsibility for personal performance.

     d. Adaptability - flexibility in handling change.

     e. Innovativeness - being comfortable with and open to novel ideas and new information.

   • Self-motivation

     a. Achievement drive - striving to improve or meet a standard of excellence.

     b. Commitment - aligning with the goals of the group or organization.

     c. Initiative - readiness to act on opportunities.

     d. Optimism - persistence in pursuing goals despite obstacles and setbacks.

2. Social competence

- Social awareness

  a. Empathy - sensing others, feelings and perspective, and taking an active interest in their concerns.

  b. Service orientation - anticipating, recognizing, and meeting customers' needs.

  c. Developing others - sensing what others need in order to develop, and bolstering their abilities.

  d. Leveraging diversity - cultivating opportunities through diverse people.

  e. Political awareness - reading a group's emotional currents and power relationships.

- Social skills

  a. Influence - wielding effective tactics for persuasion.

  b. Communication - sending clear and convincing messages.

  c. Leadership - inspiring and guiding groups and people.

  d. Change catalyst - initiating or managing change.

  e. Conflict management - negotiating and resolving disagreements.

  f. Building bonds - nurturing instrumental relationships.

  g. Collaboration and cooperation - working with others toward shared goals.

  h. Team capabilities - creating group synergy in pursuing collective goals.

* http://www.businessballs.com/emotionalintelligencecompetencies.pdf

An article chiefly developed by Cary Cherniss and Daniel Goleman, featuring 22 guidelines which represent the best current knowledge relating to the promotion of EQ in the workplace, summarize as:

A. Paving the way

- assess the organization's needs
- assessing the individual
- delivering assessments with care
- maximizing learning choice
- encouraging participation
- linking goals and personal values
- adjusting individual expectations

- assessing readiness and motivation for EQ development

B. Doing the work of change

- foster relationships between EQ trainers and learners
- self-directed change and learning
- setting goals
- breaking goals down into achievable steps
- providing opportunities for practice
- give feedback
- using experiential methods
- build in support
- use models and examples
- encourage insight and self-awareness

C. Encouraging transfer and maintenance of change (sustainable change)

- encourage application of new learning in jobs
- develop organizational culture that supports learning

D. Evaluating the change

- evaluate individual and organizational effect

2.30.1 Emotionally Positive or Negative?

Emotional intelligence is a complex interaction of several features (*e.g.* prosody, content, face, gesture, body posture). Words acquire emotional content or meaning through association with other emotional stimuli, such as visual images. Emotional intelligence is a means to personal tranquility and serenity, but such tranquility and serenity cannot coexist with self-despite, nor be maintained without communicating itself to others. This enables language (negative and positive lingual stimuli) to convey the pleasant or unpleasant nature of an object and to elicit positive and negative emotions. Not only language expresses our emotion, but also the colour of emotional responses. For example, liking ourselves and accepting our emotional behaviour does not mean giving our emotional response free rein or espousing the doctrine that emotion justifies foolish or irresponsible actions.

Let us see our responses to the following phrases:

Question	My answer	Your answer	Correct answer
1. Kids will mostly succeed if they read: a. educational books b. mathematics			

c. fairy tails	a, b, c		a, b, c
2. How would you describe your brain? a. genius b. talent c. intellect	a, c, b		c, b, a
3. Aging is: a. a merit b. a disgrace c. a dignity	a, c, b		a, b, c
4. War is: a. a sorry, inevitable consequence of beloved human nature b. a terrible crime c. when we are truly alive	b, c, a		a, c, b
5. Death is: a. not dying b. nothing at all c. only the beginning	a, c, b		c, b, a

2.32.2 Emotionally Introvert or Extrovert?

Do introverts lack emotional intelligence? Does being an introvert or extrovert have a direct impact on how emotionally intelligent you are? Is emotional difference what separates Introverts from Extroverts? Is this something that is hard to learn?

The extrovert feels recharged by spending time in the company of friends and others, the introvert by spending time isolated, but the brand of either does not hold any promise of emotional health or a life that functions better. Either can be happy or not with their need to spend time with others or alone. Introspection of whether an introvert or extrovert does not mean any thing of how well adjusted they are emotionally or how well or to their own satisfaction they function in the world.

An excess of introversion means selfishness and isolation from social interaction, and prevents creativity and effectiveness of achieving the goal. The excessive extrovert, however, discourages and oppresses brain openness, and sometimes replaces the intelligence (IQ) with a fictitious confident self-image.

Extroverts and introverts attract and repel in love all the time – they attract with friends and peers because of dominance and weakness but they repel when a strong relationship or marriage is sought because of communication differences.

Let us see our responses to the following phrases:

Question	My answer	Your answer	Correct answer

1. When you go for groceries, you buy:
 	
  a. much more than you need
  b. your current need
  c. less than you need
	a, b, c		a, b, c

2. Which most closely describes your mode in Saturdays?
 	
  a. irritable and impatient
  b. happy and cheering
  c. peaceful and determined
	c, a, b		a, b, c

3. Which most closely describes your mode in Sundays?
 	
  a. irritable and impatient
  b. happy and cheering
  c. peaceful and determined
	a, c, b		a, c, b

4. Your mentor holds a door open for you. You:
 	
  a. walk through with a thankful smile
  b. say, "no, please, after you"
  c. walk through, saying, " thank you so much"
	b, a, c		b, c, a

5. A birthday party is:
 	
  a. a woman birthday
  b. a man birthday			

c. a child birthday	c, a, b		a, c, b

2.30.3 Emotionally a Follower or a Leader?

A leader's mood has the ability to affect the organization from the top and it can positively or negatively affect the organization. The followers cannot motivate when their leaders lack the emotional intelligence to lead them effectively. Compulsive emotional leaders, like geniuses, can achieve remarkable things because they feel others who are central to their marketing and efficient management.

Naturally, a leaders' emotion, positive or negative, has a strong influence on employees' productivity (efficiency and productivity) at the workplace. When the leader exhibits positive moods or attitudes that energize employees, there will be high morale and energy. The negative attitude comes from the fact that some leaders are not dependent on the approval of their followers, and this can sentence their employees to professional, if not actual, death.

Daniel Goleman, in his ground breaking exposition on "emotional intelligence" pointed out that emotional intelligence encompasses two main domains: the personal competence and social competence. According to him, personal competence is the leaders' ability to manage himself and his emotions. This actually has two parts—self-awareness (ability to read his own emotions) and self-management (ability to control his own emotions). Social competence on the other hand, is the leaders' ability to manage relationship and other peoples' emotion. According to Goleman, this also has two parts—social awareness (ability of the leader to sense, understand, and react positively to others emotions) and, relationship management (the ability of the leader to inspire, influence, handle interpersonal interactions, and develop others).

Let us see our responses to the following phrases:

Question	My answer	Your answer	Correct answer

1. On vacation, you:
 	
    a. remain at all times in contact with the office?
    b. phone once or more a day to see the situation
    c. forget work and just devote your self to your family
	a, b, c		a, b, c

2. At weekends and holidays, you read which page of the newspaper first?
 	
  a. social/advertisement/ sports
  b. financial
  c. others
	b, c, a		b, a, c

3. Who should be working hardest and honest in the working place?
 	
  a. the boss
  b. the white collars
  c. the blue collars
	c, b, a		a, b, c

4. If one of the subordinate arrived consistently late, you would:
 	
  a. have him/her in for a stern warning
  b. ask about his/her circumstances and health
  c. not mind as long as his/her productivity was OK
	a, c, b		a, b, c

5. You read mostly:
 	
  a. books and magazines on productivity, manpower, laws and ethics, management
  b. biographies
  c. fiction
	a, b, c		a, b, c

2.30.4 Emotional Flooding

Most of the time, as adults, we can manage our emotions by processing them through the "emotional part of the brain"— the cerebral cortex. This part of our brain is responsible for self-control and judgment. In children the "emotional part of the brain" is not fully developed. Children get emotionally flooded much more easily than adults because they process their experiences through their "emotional brain"—the limbic system (The limbic system includes the hippocampus, amygdala, anterior thalamic nuclei, and limbic cortex, which suggestively support a variety of functions including emotion, behaviour, long term memory, and olfaction). This part of the brain handles emotional responding and pleasure seeking.

This is the kind of situation the noted psychologist and couples' expert John Gottman terms "emotional flooding." This term refers to relationships where aggressive and defensive reflexes have become a way of life between two or more people. These reflexes are triggered by a combination of frustrations, accumulated resentments and misunderstandings that could lead to a situation in which your dearest friend and closer companion has become your bitterest enemy.

Your perspective emotional flooding is influenced by childhood experiences: your relatives, your upbringing, your schooling, and other life experiences.

Let us see our responses to the following phrases:

Question	My answer	Your answer	Correct answer

1. You are:
 	
  a. understanding, tolerant, forgiving
  b. devoted, passionate, courteous
  c. impossible, but you love him/her
	b, a, c		c, b, a

2. Your wife is:
 	
  a. always difficult and demanding
  b. always easy, satisfied
  c. unworkable
	c, a, b		c, b, a

3. Your husband is:
 	
  a. mystifying, hard to understand
  b. selfish, careless
  c. inconsistent, moody
	c, b, a		c, a, b

4. My friend is mean, unwise, weird
 	
  a. always
  b. often
  c. never
	b, a, c		b, c, a

5. He/she refused your idea, you say:
 	
  a. "don't worry, forget the idea"
  b. "this is you habit, you never accept any idea"			

c. "I shall ask some body else"	a, c, b		a, c, b

2.30.5 Do You Know Yourself Well? Do You Make the Most of Yourself at Work?

If you know yourself well, you've got a better chance of understanding what opportunities to capture and which to avoid. If you know yourself well, you're more likely to understand others and get along with them better. In other words, if you know yourself, you're more likely to succeed. Training and learning are very important to narture yourself at work. The more you learn how to fine tune, improve and enable yourself, the better you'll perform at your job and in every other area of your life.

To succeed in your job, psychologists recognize four basic personality styles: analytical, amiable, expressive, and driver. Usually, each of us exhibits personality characteristics unique to one of the styles. However, we also possess characteristics to a lesser degree in the other styles. To determine your unique style, you can take a Myers-Briggs assessment, or other recognized assessments. The assessment is necessary for those who can make the most for themselves at work, those who can win success and those who can enjoy their work.

Let us see our responses to the following phrases:

Question	My answer	Your answer	Correct answer
1. How many hours of leisure do you have in every twenty-four hours? a. 4-6 b. 7-9 c. 10-16	a, b, c		c, b, a
2. Some yawns whilst you are delivering a speech. Do you: a. assume that he/she is not interested or committed? b. Assume that you are being boring? c. Not link the two things at all?	b, a, c		c, b, a
3. You are negotiating a contract. You approach the negotiations: a. ready to negotiate			

	a, c, b		b, a, c
b. ready to accept the lowest price c. ready to share the contract			
4. An employee talks too much in a workplace: As a supervisor, do you: a. warn him? b. demand that he stops talking? c. express your feeling directly to him?	c, b, a		c, b, a
5. You are in what seems to you a dead-end job, unnoticed, your work taken for granted. Do You: a. enroll in a training course, motivate your colleagues, and then if you see no further prospects, look for a new job? b. look for a new job? c. complain to you supervisors, then, if you see no improvement, threaten to look for a new job?			

2.30.6 Emotionally Independent

One of the most essential qualities of leadership is emotional independence. Independence is about being flexible and adapting to changing situations. Independence is about being just even in the face of strong conflicting emotion; independence is about accepting that you are responsible for anything and everything in your life and that you can change that whenever you wish. Independence is about trusting yourself, and believing in yourself. Independence is about being able to go on regardless of another's opinion, regardless of emotional lyrics, and regardless of other's life style.

Emotionally independent people recognize their own pattern, recognize that their feelings are valid, not just those around them, and stay away from people who are emotional predators.

Let us see our responses to the following phrases:

Question	My answer	Your answer	Correct answer

1. Sex is for you:
 	
  a. a way of asserting yourself and gaining pleasure
  b. a way of surrendering yourself and giving pleasure
  c. a way of attaining all of these			

	a, c, b		a, c, b

2. A TV is:
 	
  a. necessity
  b. basic
  c. a part of the furniture			

	a, b, c		b, a, c

3. You receive a bill on the15th of the month. Do you:
 	
  a. pay the bill on date?
  b. don't open the bill until the end of the month?
  c. leave it unopened until you have the funds?			

	a, b, c		a, c, b

4. Your telephone rings at 2:00 am. Do you?
 	
  a. answer it?
  b. ask your wife to answer?
  c. leave it ringing?			

	b, a, c		a, b, c

5. Your previous boss invited you on a party at his home. You are:
 	
  a. thrilled to see your old friends
  b. ostensibly cheering
  c. declining			

	a, c, b		b, c, a

2.30.7 Emotionally Knowledgeable

As a result of globalization, there is a great demand for global effective leaders who are adequately, emotionally and culturally knowledgeable (CQ). Such leaders need to possess the required knowledge measured in terms of EQ and

CQ. However the real "supply" of contemporary managers and emerging global leaders equipped in high EQ and CQ is seriously very low.

Emotional knowledge is not only about gaining the ability to recognize and positively manage emotions in you, in others and in groups. It is also about emotion where suitable. Recent studies indicate that emotional knowledge is a powerful key to effective leadership. Emotional knowledge is a valuable tool you need to ensure that you are a strong, emotionally intelligent leader, and to gain authority and success you strive to be a leader.

Emotional knowledge and literacy can be learned to develop cohesive, emotionally intelligent teams, to monitor and adjust your emotions and behaviours for yourself and your group, to gain an honest and accurate awareness of yourself and others, and to understand and apply the psychology of leadership.

Let us see our responses to the following phrases:

Question	My answer	Your answer	Correct answer

1. You enraged with your friend. Do you:
  a. refuse to speak until he speaks to you firstly?
  b. go for a walk?
  c. Go and say sorry to him?
	c, b, a		b, a, c

2. When you show resentment. Do you:
 	
  a. expect a stimulus to change?
  b. expect to gain sympathy from others?
  c. Expect to hurt yourself?
	a, b, c		a, b, c

3. you are outraged by the speech of your boss. Do you:
 	
  a. leave the area where the speech is given?
  b. you stay but with resentment?
  c. master you emotion and tell him why this was not appropriate?
	b, c, a		c, b, a

4. Your child, in direct contravention to your orders, runs into the busy road. You pull			

him/her back and: a. smack or shake him/her b. master your emotion and explain why this was not a good idea c. master your emotions and resolve upon subsequent punishment	b,c,a		a, b, c
5. You (young person) like music to be: a. classical b. rock'n'roll c. metal	a, b, c		b, c, a

2.30.8 Emotional Energy

Emotional energy is the bond between our minds and our energy, and how it all relates to health. Our soul communicates with us through emotional energy vibrations. A person's total energy is how ready, willing, and able he/she is to live -- to take on challenges, to fight for what's important to him/her, to work to make his/her dreams come true, to care for loved ones, to build a better future, to enjoy life to the fullest in the present. Emotional energy is something that we feel in our heart/our gut, something that resonates within us. Only 30% of the energy required for this is physical. That is, only 30% of the energy necessary for this comes from physical sources like rest, diet, exercise, and health. New research shows that 70% of total energy comes from nonphysical sources, what we call emotional energy. You have emotional energy when you feel up for your life -- hopeful, positive, engaged, charitable, caring, patient, focused, and loving, (Mira Kirshenbaum's latest book).

Emotional energy cannot be measured or weighed by an X-ray or MRI ("The MRI shows you have got 2 pounds of happiness or grief in your body"), thus emotional energy is discounted and devalued. It is like latent energy stored in our body in a pressurized, explosive state, as a result of being held back and not released. Such an explosive state, if not released properly, it would transform to a disease such as cancer, depression, or emotional wounds such as obsession and compulsion. Because of our unhealed emotional wounds, the emotional truth cannot match the original intuitive emotional truth born with us. That is why we receive and perceive negative messages about emotional expression and feelings.

After all, your emotional energy is what keeps you going during times of distress and peace. It is the driving forces of the psyche, emotionally as well as intellectually. It's also what enables you to take care of yourself and other people.

There are five types of energy which are quite distinct. They are absolutely interrelated and operate on the gestalt principle, constituting the personality of the individual:

a. Physical energy, through which the body is able to generate energy for autonomous functions, activities and behaviors.
b. Mental energy through which the brain is able to perform thinking, reasoning and solving of problems.
c. Psychic energy, which is referred to enthusiasm, drive, and resilience.
d. Sexual energy, which is referred to the primeval urge to procreate.
e. Religious or spiritual energy, which is referred to as the purest form of energy and creates an environment of vibrancy. It is the 'spirit' or the breath of life.

Let us see our responses to the following phrases:

Question	My answer	Your answer	Correct answer
1. when you face a significant defeat or rejection, do you: a. feel discouraged? b. feel threatened by feelings of sadness, anxiety, annoyance? c. feel that you cannot cope?	a, c, b		a, b, c
2. When you go to work, do you feel: a. dreadful b. energetic c. resignation	b, c, a		b, c, a
3. You're emotionally exhausted, you are: a. dominated by strong negative feelings b. no longer do most of the things you used to do that gave you pleasure c. faced with just one more big difficulty, you won't be able to cope.			
4. How do you increase your emotional energy? a. by eliminating the painful, limiting emotions and stress that prevent you			

from performing at your best b. by marshaling emotions in service to a goal is essential for paying attention, for self-motivation and mastery and for creativity c. by getting plenty of healthful, nutritious food so you can keep your physical energy at high levels		a, b, c			b, a, c
5. To maintain Positive Attitudes, one should: a. maintain a high energy level b. get rid of Worry, anxiety, fear, anger and depression c. scan your psychological state to see what you are thinking and feeling		c, a, b			b, a, c

2.31 Cultural Intelligence

Cultural intelligence and cultural quotient (CQ), is a theory within management and organizational psychology, positing that understanding the impact of an individual's cultural background on their behaviour is essential for effective business, and measuring an individual's ability to engage successfully in any environment or social setting. Cultural intelligence is your capability to grow personally through continuous learning and good understanding of diverse cultural heritage, wisdom and values, and to deal effectively with people from different cultural backgrounds and understandings.

CQ has been gaining acceptance throughout the business community. CQ teaches strategies to improve cultural perception in order to distinguish behaviours driven by culture from those specific to an individual, suggesting that allowing knowledge and appreciation of the difference to guide responses results in better business practice.

CQ is developed through:

- cognitive means: the head (learning about your own and other cultures, and cultural diversity)
- physical means: the body (using your senses and adapting your movements and body language to blend in)

- motivational means: the emotions (gaining rewards and strength from acceptance and success)

CQ is measured on a scale, similar to that used to measure an individual's IQ. People with higher CQ's are regarded as more able to successfully blend in to any environment, using more effective business practices, than those with a lower CQ.

Cultural intelligence will help you manage effectively cross-cultural differences, in particular:

1. It will reduce the cultural barriers between 'us' and 'them' and allow you to predict what 'they' are thinking and how they will react to your behavior patterns. The goal is to mingle their culture with yours
2. It will harness and enhance the power of the cultural diversity of beliefs, values, perceptions, expectations, attitudes and assumptions.

In today's globalizing world, cultural intelligence is an important tool for every manager who deals with diverse teams of employees, customers, partners, competitors, government, and other business players.

"The world is becoming more interdependent; to keep pace we must all learn to think globally - we must all develop our cultural intelligence", write intercultural relations experts David C. Thomas and Kerr Inkson in their essential guide to global cultural understanding *Cultural Intelligenec, Second Edition: Living and Working Globally.*

2.31.1 Cultural Intelligence and Globalization of Business

People in this world are all having increasingly global lives, and managers, in particular, have businesses that are more global than most. In the beginning of this century, we are seeing and understanding the importance of cultural intelligence in conquering obstacles of globalization of business. Globalization means an increase in the permeability of traditional barriers, including those around countries (even inside one country), economies, industries, organizations and institutions. Globalization has recently been accelerated by many factors in the international trade environment, including the following:

- New international trade and business agreements
- The increase of international trade
- The growth of multinational cooperations
- The privatization of state enterprises
- The ability to relocate businesses, particularly manufacturing and industrial, where cost and taxes is lowest
- The restructuring and down sizing of enterprises

- The ease of financial transactions between banking institutions
- The establishing of free trade agreements between countries
- The expansion of international migration
- The ease of using information and communication technology to transcend time and distance

Globalization business is growing constantly because the world is now a much smaller place than ever before and the number of intercultural business and personal encounters is developing continually. All too often, people lacking sensitivity cultural differences expect others to be more like them, resulting in contacts and understanding in a better mutual interest.

2.31.2 Cultural Intelligence and Corporate Culture

Corporate culture requires that individuals to be sensitive to different cultures. To do all this, individuals, whether they are at home or abroad, need cultural intelligence. It is needed to manage the stress of culture shock and the consequent frustration and confusion that typically results from clashes of cultural differences. It is essential in facilitating effective cross-cultural adjustment.

Companies adopt a broad range of solutions to develop intercultural competences. Solutions include face to face training, eTraining, performance improvement workshops, individual coaching (for senior people or expatriates), videos, etc. Coaching must follow positive behaviours, must have a mind-set that is positive and solution-oriented, and must be proactive and challenges negative behaviour displayed by others.

Here are some ways corporate professionals are applying cultural intelligence:

1. Enhancing their own CQ in order to work with employees coming from a variety of backgrounds.
2. Drawing upon CQ to analyze various jobs within the organization and identifying which ones require the strongest degree of cultural intelligence.
3. Altering performance appraisals in light of the cultural backgrounds of various personnel (e.g. giving managers tips on how to offer criticism to someone from an Asian background as compared to someone from a Latin background).
4. Creating policies that respect cultural differences while still remaining true to the corporate culture and brand (e.g. head coverings for women).
5. Offering cultural intelligence training and consulting employees as part of their professional development.
6. Prioritizing cultural intelligence among all new hires by assessing it through a cultural intelligence inventory, finding better ways to motivate people to participate in diversity training.
7. Drawing upon CQ to help bridge the generational divide that exists among many subcultures of an organization.

8. Developing training around the four capabilities of CQ (Drive, Knowledge, Strategy, and Action).
9. Assessing current and future personnel by interviewing them in light of the four capabilities of CQ.

2.31.3 Cultural intelligence and Global Inequality

Cultural diversity is differences (in race, ethnicity, language, nationality, religion, socioeconomic, etc) among various groups within a community. A community is said to be culturally diverse if its residents include members of different groups.

Cultural diversity is becoming more and more important.

- White males occupy 5% fewer management jobs in 2006 than in 1998, and every other racial/gender group occupy more management jobs.
- The U.S. population of foreign born residents is 12.4%, an amount of international diversity that the U.S. has not seen since 1920.
- 90% of leading executives from 68 countries named cross cultural leadership as the top management challenge of the next century.
- The proportion of revenue coming from overseas markets is expected to jump by an average of 30 to 50 percent over the next 3 to 5 years

The following factors are indicators affecting the cultural intelligence:

- Language - Multi-International teams have individuals who will not be able to converse in their native tongue. Undoubtedly, this will lead to some form of misinterpretation when working as a team for at least one member of the group.
- Language style - Varying nationalities communicate differently. When working in teams some individuals may be very direct and blunt, while others may be indirect and vague in their communication. This is especially evident when we compare American English and British English.
- Culture – It underlies invisible values and beliefs that make up behaviors that are unique to each society. For example, attitudes toward time, different value systems, and preconceptions & stereotypes may have different approaches and tasks.
- Income per capita – This will affect the economical intelligence and wellbeing.
- Health – It affects life expectancy and CQ (CQ is the product of IQ and EQ).
- Political stability and security – Cultural intelligence provides a baseline for designing successful strategies to interact with foreign peoples whether they are neutrals, people of an occupied territory, or enemies,
- Family life – A poor family may have a more generous spirit than a rich one, and this is a part of the CQ.

157

- Community life – adjusting to a new culture and work environment is a difficult process and can generate a variety of adverse reactions and feelings.
- Climate and geography – Knowledge of climate and geography provide means to comprehend the rapidly changing physical and cultural environments of the world, and thus, prepare us to be better global citizens.
- Political freedom and civil liberties – Good correlation between both education and life satisfaction and political freedom.
- Gender equality – There is no indication that gender influences cultural intelligence.
- Job security – A sense of job security is an indicator to experience personal and professional growth.

2.31.4 Comparison of Global Cultural Intelligence

We live and work in a world that is an integrated entity, increasingly influenced by external cultural factors. Although it has never been disputed that there are systematic differences between average scores in IQ tests (IQ +EQ = CQ) of different population groups, sometimes called "racial IQ gaps", there has been no agreement on whether this is mainly due to environmental and cultural factors, or whether some inherent hereditarian factor is at play, related to genetics.

There are observed differences in average test score achievement between racial groups, which vary depending on the populations studied and the type of tests used. In the United States, self-identified Blacks and Whites have been the subjects of the greatest number of studies. The Black-White IQ difference is largest on those tests that best represent the general intelligence factor g, Figure (2.19).

Figure (2.19): Race difference in intelligence

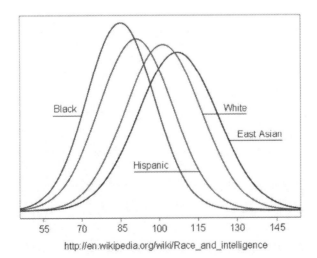

http://en.wikipedia.org/wiki/Race_and_intelligence

158

In today's globalizing world, cultural intelligence is a necessary tool for every manager who deals with diverse groups of employees, customers, partners, competitors, government, and other business players. Figure (2.20) show comparison of management and leadership (components of CQ) in the five continents.

Figure (2.20): Comparison of management, leadership and cultural intelligence in the five continents (Asia is represented by Asia-Pacific)

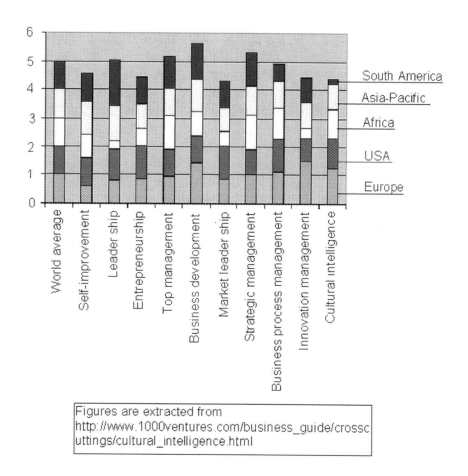

Figures are extracted from http://www.1000ventures.com/business_guide/crossc uttings/cultural_intelligence.html

2.32 Environment and Intelligence

Neurons and cognitive scientists suggested that different intellectual abilities are based on a neural network that requires environmental stimulation for development -- and are open to change. However, intelligence researchers argue that there is a general factor of intelligence g that is highly heritable and defines intelligence as an overall innate ability to perform well on different measures of intelligence -- which are not open to change.

Environmentalists who support the idea that the environment has a large effect on intelligence performed many experiments to prove that environmental stimulation affects cognition and intellectual development from a neurobiological perspective. In an experiment, four different habitats were set up to test how environmental enrichment or relative impoverishment affected rats' performance on various measures of intelligent behavior. First, rats were isolated, each to its own cage. In the second condition, the rats were still in isolation, but this time they had some toy, or enriching object in the cage with them. The third condition placed the rats in cages with each other, so they were receiving social enrichment, without any enriching toys. The fourth and final condition exposed the rats to both social interaction and some form of object enrichment. In measuring intellectual capacity, the rats that had both forms of enrichment performed best, the ones with social enrichment performed second best, and the ones with a toy in their cage performed still better than the rats with no toy or other rats. When the volume of the rat's cortices was measured the amount of enrichment again correlated with larger volume, which is an indicator of more synaptic connections, and greater intelligence, MC Diamond et al. Effects of environmental enrichment and impoverishment on rat cerebral cortex. Journal of Neurobiology. 1972; 3(10): 47-64.

Researchers found that reasoning capacities are a function of the connections of the neural system, connections that develop during childhood as the neural system adapts to the environment. However, they also note that the capacity of the brain to adapt its connections to environmental stimuli diminishes over time, and therefore, it would follow that there is a critical period for intellectual development as well (usually 16 years). While the critical period for the visual cortex ends in early childhood, other cortical areas and abilities have a critical period that lasts up through maturity (age 16); the same time frame for the development of fluid intelligence.This corresponds to the observation that reasoning capacities are observed to no longer increase after time.

Environmental factors such as moral, political, ethical, educational, and social, physiological could contribute to individual differences in intelligence.

2.32.1 Genetic-Environmental Debates

Galton in his book "Hereditary Genius: Its Laws and Consequences" (1869) had observed that the gifted individuals tended to come from families which had other gifted individuals. He went on to analyze biographical dictionaries and encyclopedias, and became convinced that talent in science, the professions, and the arts, ran in families.

Galton promoted his theory one step further, to suggest that it would be "quite practicable to produce a high gifted race (eugenics) of men by careful marriages during several consecutive generations".

His suggestion has had massive social consequences and has been used to support apartheid policies, sterilization programs, and other acts of withholding political freedom, civil liberties and basic human rights from minority groups.

Massive research on intelligence took place after Word War I to conclude that the Galton theory was erroneous and immoral. Researchers found blacks from Illinois had higher IQ scores than whites from 9 southern states - a finding difficult to reconcile with the simple idea that whites are intellectually superior to blacks. As a result, a number of psychologists in the 1920s and 1930s shifted their position towards the environmental camp and retreated form the genetic camp. To boost the environmental effect on intelligence, efforts were made to improve poor educational achievement through special schooling, and to alleviate poor living conditions through welfare programs.

Then Bell Curve by Herrnstein and Murray (1994) came to reconcile the dispute between genetical and environmental supporters. The Bell Curve shows that no less than 40% and no more than 80% of is inherited genetically from parents. In other words, environmental effect on intelligence has the average between 20% and 60%.

John B. Watson, 1924 wrote "Give me a dozen healthy infants & my own specific world to bring them up in, & I'll guarantee to take any one at random & train him to become any type of specialist I might select - doctor, lawyer, artist, merchant, chef & yes, even beggar & thief, regardless of his talents, penchants, tendencies, abilities, vocations, and race of his ancestors."

The Flynn effect proved that it is the environmental effect and not the heredity effect. In the 1980s, an NZ-based political scientist, James Flynn, noticed that IQ was increasing in all countries all the time, at an average rate of about 3 IQ points per decade i.e. the average IQ across the world has risen over 1 standard deviation (i.e. 15 points) since WWII - predominantly due to environmental effects. As a result, new norms continue to be used to rescale IQ tests to '100', Figure (2.21).

Figure (2.21): Intelligence rising per time

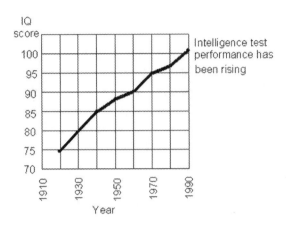

2.32.2 Factors Affecting Intelligence (Other Than Genetics)

Bouchard & Segal; Liungman, 1975 reported that many factors could attribute to intelligence. These factors vary to a greater or lesser extent in relation with IQ. However, not all of these factors support an environmental view. These factors are:

- Infant malnutrition
- Birth height and weight
- Birth number and order
- Number of siblings
- Number of schooling years
- Social group of parental home
- Father's profession
- Father's economic status
- Degree of parental rigidity
- Parental ambition
- Mother's education
- Hours of TV viewing
- Average book-reading
- Self-confidence according to attitude scale measurement
- Age (negative relationship, applies only in adulthood)
- Degree of authority in parental home
- Alcoholism
- Criminality
- Mental disease
- Emotional adaptation

162

2.32.3 Braitenberg Vehicle (Environmental Representation)

In his book *Vehicles: Experiments in Synthetic Psychology*], Valentino Braitenberg (an Italian Austrian phsychologist) describes some experiments involving little machines, which are made up of the following parts:

- 2 wheels – one on the left, and one on the right
- A number of sensors mounted on the front of the vehicle which respond to elements in the environment (heat, light, sound, pressure, etc)
- Connections from the sensors to the motors - these connections either interrupt or supply power to the motors.

At the heart of the Braitenberg vehicles is a description of a basic neuron setup. At the front end we find a sensor. In this illustration the sensor detects the intensity of light and outputs a proportional signal. Consider the sensor connected to the neuron as modular and interchangeable. Other sensors can be incorporated to detect any number of environmental components such as heat, pressure, sound, vibration, magnetic fields (compass), electrical fields, radioactivity, gases (toxic or otherwise), etc.

If the sensor is for light detection, the vehicle will stimulate its motion accordingly. A first agent has one light-detecting sensor that directly stimulates its single wheel, implementing the following rules:

- More light produces faster movement.
- Less light produces slower movement.
- Darkness produces standstill.

This behavior is simulated as an environment and its effect on a creature (like a cockroach) that is afraid of the light and that makes it move fast to get away from it. Its goal is to find a dark spot to hide.

2.32.4 Design Principle of Autonomous Agents

New approaches to artificial intelligence spring from the idea that intelligence emerges as much from physical embodiment like cells, bodies, and societies as it does from evolution, development, and learning. Traditionally, artificial intelligence has been concerned with reproducing the abilities of human brains; newer approaches take inspiration from a wider range of biological structures that that are capable of autonomous self-organization. It has concentrated predominantly on lower forms of biological life as an inspiration for the design of robot control systems by exploring reactive and hybrid behavior-based systems, evolutionary and reinforcement learning methods, and enabling perceptual paradigms such as active vision and task-oriented perception. Examples of these new approaches include evolutionary computation and evolutionary electronics, artificial neural networks, immune systems, biorobotics, and swarm intelligence—to mention only a few. Contributions are encouraged from across the broad range of disciplines-computer science, engineering, the biosciences (neuroscience, psychology, and ethology), organizational behavior, and economics-that are contributing to progress in these areas. Computer science has emerged in the production of robotic systems that draw heavily on biology and ethology; it uses the tools of neural networks, genetic algorithms, dynamic systems, and biomorphic engineering.

2.32.5 The Principle of Parallel, Loosely Coupled Processes

The feasibility of parallel (simultaneous) processing can be demonstrated by neurons in the brain. Aggregate speed with which complex calculations are carried out and transmitted by neurons is tremendously high, even though individual responses of neurons are too slow (in terms of milliseconds). Parallel processors use divide and conquer algorithm, which is based on multi-branched recursion and breaking down a problem into two or more sub-problems of the same (or related) type, until these become simple enough to be solved directly.

Parallel processing provides cost-effective solutions to communications and speed by increasing the number of CPUs (in computers) or converters and nodes in artificial robots. This results in much higher computing power and performance than could be achieved with traditional computing systems.

The objective is to build a brain and bodies with artificial neural network. It includes:

- Embedded artificial neural network
- Simulation of different types of neurons
- Modeling neurons and neuron networks
- Input and output (analog and digital)
- Internal box includes function initiations and activations
- Representation of nodal matrices
- Compilers
- Motor action includes obstacles, collision and feed back loops for correction
- Agent-environmental interaction
- Redundant circuitries to boost and correct malfunctioning
- The output is the social interaction robot

The principle of the parallel, loosely coupled processes is that intelligence is emergent from a large number of parallel, loosely coupled processes, as pointed out above, to interact with the environment through agents embedded in the machine (robot). These processes run asynchronously and are coupled to the agent's sensory-motor apparatus. The design does not require a centralized brainpower with lower level design (insects), but in higher level design (human) it may need a little centralized brain. Generally, the principle of the parallel, loosely coupled processes contrasts with classical thinking where a centralized center of controlling intelligence is there. Parallel processes do not contest with classical intelligence except that the later has a central integration (brain).

2.32.6 Environmental Influences on Development of Intelligence

Research on factors that cause or influence the development of intelligence from many perspectives can be understood within an evolutionary framework in which organism and environment combine to produce development. Species-normal genes and environments and individual variations in genes and environments both affect personality, psychopathology, cognition, social, and intellectual development. These domains are used as examples to integrate theories of normal development and individual differences. Within the usual samples of European, North American, and developed Asian countries, the results of family and twin studies show that environments within the normal species range are crucial to normal development.

One of the most important findings that have emerged from human behavioral genetics involves the environment rather than heredity, providing the best available evidence for the importance of environmental influences on development of intelligence.

In a study by Eric Turkheimer, University of Virginia, a substantial proportion of the twins were raised in families living near or below the poverty level. Biometric analyses were conducted using models allowing for components attributable to the additive effects of genotype, shared environment, and non-shared environment to interact with socioeconomic status (SES) measured as a continuous variable. Results demonstrate that the proportions of IQ variance attributable to genes and environment vary nonlinearly with SES. The models suggest that in impoverished families, 60% of the variance in IQ is accounted for by the shared environment, and the contribution of genes is close to zero; in affluent families, the result is almost exactly the reverse.

Deprived or abnormal rearing conditions induce severe disturbance in all aspects of social and emotional functioning, and effect the growth and survival of dendrites, axons, synapses, interneurons, neurons, and glia. The amygdala, cingulate, and septal nuclei develop at different rates which correlate with the emergence of wariness, fear, selective attachments, play behavior, and the oral and phallic stages of development. These immature limbic nuclei are "experience-expectant," and may be differentially injured depending on the age at which they suffer deprivation.

Other factors such as parental education also affect the level of intelligence of children. Both heritability and the shared environmental estimate were moderated, however, by the level of parental education. Specifically, among more highly educated families, the level of heritability is increased and conversely, among less well-educated families, heritability decreased and shared environmental influences increased.

The nature-nature interplay influences the development of the three most common psychological problems in childhood: communication disorders, mild mental impairment and behavior problems. Results of a relatively large number of studies have shown that problem behavior may be functionally related to and reinforced by events in the social environment. A basic assumption is that problem behavior is the final outcome of a learning process: this type of behavior may be related to antecedent environmental impact as well as consequent environmental stimuli. Consequent events may positively and negatively reinforce the occurrence of problem behavior.

2.33 Brain and Environmental Richness (Neural)

Brains manage their constituent cells in a way unlike other human cells and tissues. Most cells and tissues replenish themselves because their cells reproduce. But the cells in a brain steadily decline in number during the brain's lifetime. This may be because age-related changes in brain white matter have taken precedence in explaining the steady decline in cognitive domains seen in non-diseased elderly. Unlike other types of cells, brain cells, or neurons don't replicate, but like other cells, they do die. And yet, with its neuronal population steadily declining, a human brain grows dramatically in weight and volume during its early years. Brain researchers account for this paradox by pointing out that, although a brain has fewer neuron cells as time passes, the cells that remain continue to grow by burgeoning rich networks of connections to their neighbors.

A baby's brain grows quickly in the uterus, and after birth it continues to grow but not by creating more cells. Once the baby is born, a brain grows by creating more connections among its cells. The density of synapses in the brain tissue peaks in humans between the ages of three and six, then tapers off, by about 50 percent, to adult levels by late adolescence. The human brain reaches about 95 percent of its adult volume typically by the age of five.

It is demonstrated that the trajectory of change in the thickness of the cerebral cortex, rather than cortical thickness itself, is most closely related to level of the IQ. Using a longitudinal design, we find a marked developmental shift from a predominantly negative correlation between intelligence and cortical thickness in early childhood, to a positive correlation in late childhood and beyond. Additionally, the levels of intelligence is associated with the trajectory of cortical development, primarily in frontal regions implicated in the maturation of intelligent activity, Booth, J. R. et al. Neural development of selective attention and response inhibition. Neuroimage 20, 737–751 (2003). More intelligent children demonstrate a particularly plastic cortex, with an initial accelerated and prolonged phase of cortical increase, which yields to equally vigorous cortical thinning by early adolescence. This study indicates that the neuroanatomical expression of intelligence in children is dynamic. Structural neuroimaging studies generally report a modest correlation, Sowell, E. R. et al. Longitudinal mapping of cortical thickness and brain growth in normal children. J. Neurosci. 24, 8223–8231 (2004).

Researchers from neuroscience, animal behavior, ecology and evolutionary biology, informatics, and physics examined how to adapt cortical thickness and synapse formation in childhood for the purpose of:

- Increasing the respond of the nervous system to environmental stimuli
- Using some types of stimuli to alter genes expression in the brain
- Contributing to genetic diversity through common environmental stimuli
- Gaining a fuller appreciation of how the brain structure control behavior and intelligence
- Studying the neuronal changes that occur in response to complex stimulation by an enriched environment

Numerous cognitive theories about how environmental enrichment affects the brain have been proposed. Among them are the arousal hypotheses (Walsh, R. N. & Cummins, R. A. Mechanisms mediating the production of environmentally induced brain changes). Psychol Bull. *82*, 986–1000 (1975) which emphasizes the so-called 'arousal response' of animals when confronted with novelty and environmental complexity, and the 'learning and memory' hypothesis (Rosenzweig, M. R. & Bennett, E. L. Psychobiology of plasticity: effects of training and experience on brain and behavior). Behav. Brain Res. 78, 57–65, 1996, in which the mediator of the morphological changes is seen in the cellular mechanisms underlying learning processes. Although the learning-and-memory hypothesis is favoured by many investigators, it is difficult to prove that the neural consequences of the enriched environment are related to learning rather than to increased voluntary motor behaviour.

To conclude, the environmental enrichment has been reported to be accompanied by improvement in cognitive performance, and has shown to accelerate development of the sensory system. Researcher's new findings highlight the potential of environmental enrichment as a promising non-invasive strategy to ameliorate deficits in the maturation of the nervous system and to promote the recovery of normal sensory functions in pathological conditions affecting the adult brain.

A new study shows that environmental enrichment increases histone acetylation which has the ability to form new memories and re-establish access to remote memories even in the presence of brain degeneration. Thus, environmental enrichment represents a promising target for therapeutical intervention in neurodegenerative diseases.

2.34 Brain Plasticity

Brain plasticity (also referred to as neuroplasticity or cortical re-mapping) is the changing of neurons, the organization of their networks, and their function via new experiences and environmental enrichment. Neuroplasticity or brain plasticity refers to the brain's ability to change and modify throughout life. The brain has the amazing ability to reorganize itself by forming new connections between brain cells (neurons). Neuroplasticity occurs in the brain at the beginning of life: when the immature brain organizes itself, in case of brain injury: to compensate for lost functions or maximize remaining functions; and through adulthood: whenever something new is learned and memorized.

According to the theory of neuroplasticity, sensing, thinking, learning, and acting actually change both the brain's physical structure (anatomy) and functional organization (physiology) from top to bottom. Neuroscientists are presently engaged in a reconciliation of critical period studies demonstrating the immutability of the brain after development with the new findings on neuroplasticity, which reveal the mutability of both structural and functional aspects. Bach-y-Rita (deceased in 2006, was the "father of sensory substitution and brain plasticity") explained plasticity by saying, "If you are driving from here to Milwaukee and the main bridge goes out, first you are paralyzed. Then you take old secondary roads through the farmland. Then you use these roads more; you find shorter paths to use to get where you want to go, and you start to get there faster. These "secondary" neural pathways are "unmasked" or exposed and strengthened as they are used. The "unmasking" process is generally thought to be one of the principal ways in which the plastic brain reorganizes itself."

It is known that normal cells in the body are replaced if they are damaged or dead. The brain cells (nerve cells) do not have this ability to replace themselves if they are damaged or dead. Each person was born with a certain number of neuron cells, the number of these cells decreased continuously, but they could continue making new connections with one another until the time he died. However, recent studies indicate that certain parts of the brains of primates, including humans, may maintain their ability to produce new neurons throughout adult life. The studies found that the stem cells found in the hippocampus divide in two. One of the two cells will remain a stem cell, while the other will turn into a neuron.

When stem cells divide, some become blood cells, muscle cells, skin cells, or functional nerve cells, with dendrites and axons that connect them to other parts of the hippocampus, a process known as neurogenesis, Figure (2.22).

Figure (2.22): Division of a stem cell into a stem cell and nerve cell

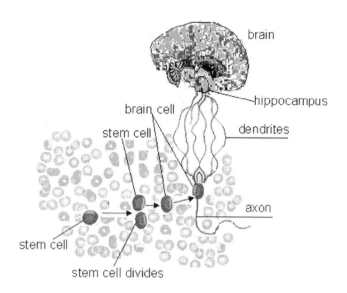

2.34.1 Facts and Stories of the Brain That Changes Itself

1. When people get older their ability to comprehend rapid speech goes down. The key idea is that if you can improve hearing comprehension, memory improves. This is because of the well-known fact that richer signals leave stronger memories.

2. At three years of age, Nico of Spain, was about to undergo a hemispherectomy. He was having close to one hundred epileptic seizures a day. The right hemisphere was the source of the trouble and surgeons, in an attempt to save his life, removed it. He is now epilepsy-free, has many friends, and with the help of a computer he is powering through school. Nico demonstrates a high IQ and possesses exceptional language skills for his age. Speech and language are understood to be primary capacities of the left brain. He also has difficulty controlling the left side of his body and he cannot see out of his left eye. After all, the visual cortex is missing for his left eye. Beyond this it is impossible to tell that Nico has only half of his brain.

3. On September 13, 1848, Phineas Gage was a railroad construction foreman in charge of blasting gang on the Rutland & Burlington Railroad outside the town of Cavendish, Vermont. After an assistant drilled a hole in a large granite rock, added gunpowder, and sand (the sand may have been omitted by accident), Gage was to add a fuse, and tamp the mixture down just prior to blasting. Gage was standing (or sitting [accounts vary]) and was momentarily distracted. No witness was able to state what the distraction was. The tamper slipped from Gage's hand and fell into the hole. A spark from the granite ignited the powder and it exploded sending the iron rod (see photograph) flying. It entered the left side of his face, shattered his upper jaw, passed behind his left eye, and

out the top of his head. The rod ended up roughly eighty feet away and Phineas seemed none the worse for the accident. He spoke to coworkers and walked under his own power to a cart that took him the three quarters of a mile to town.

Dr. Edward H. Williams examined Phineas and stated that "I first noticed the wound upon the head before I alighted from my carriage, the pulsations of the brain being very distinct. Mr. Gage, during the time I was examining this wound, was relating the manner in which he was injured to the bystanders. I did not believe Mr. Gage's statement at that time, but thought he was deceived. Mr. Gage persisted in saying that the bar went through his head....Mr. G. got up and vomited; the effort of vomiting pressed out about half a teacupful of the brain, which fell upon the floor."

Dr. John Martin Harlow was the next physician to see him. Phineas underwent a long recovery with possibly two infections, one of which left him in delirium, but by April 1849 he seemed to have completely recovered. However, according to a medical paper written two decades later, friends reported that he could no longer control his emotions or temper and some even went so far as to say Phineas was no longer "Phineas."

In 1868 Dr. Harlow wrote: "The equilibrium or balance, so to speak, between his intellectual faculties and animal propensities, seems to have been destroyed. He is fitful, irreverent, indulging at times in the grossest profanity (which was not previously his custom), manifesting but little deference for his fellows, impatient of restraint or advice when it conflicts with his desires, at times pertinaciously obstinate, yet capricious and vacillating, devising many plans of future operations, which are no sooner arranged than they are abandoned in turn for others appearing more feasible. A child in his intellectual capacity and manifestations, he has the animal passions of a strong man. Previous to his injury, although untrained in the schools, he possessed a well-balanced mind, and was looked upon by those who knew him as a shrewd, smart businessman, very energetic and persistent in executing all his plans of operation. In this regard his mind was radically changed, so decidedly that his friends and acquaintances said he was 'no longer Gage."

It is known that Gage took advantage of his notoriety and appeared at P.T. Barnum's American Museum. Later in life Gage took a job in Chile as a long distance coach driver for seven years. Prior to that he was at a livery in San Francisco for a year and a half. This indicates that Gage somehow recovered from all of the injuries and lived a socially adjusted, normal life, http://hubpages.com/hub/More-Brain-Plasticity-Science (By Liam Bean).

4. To test the idea, Alessandro Farné of the University of Claude Bernard in Lyon, France, and colleagues attached a mechanical grabber to the

arms of 14 volunteers. The modified subjects then used the grabber to pick up out-of-reach objects. Shortly afterwards, the volunteers perceived touches on their elbow and fingertips as further apart than they really were, and took longer to point to or grasp objects with their hand than prior to using the tool.

The explanation, say the team, is that their brain had adjusted the brain areas that normally control the arm to account for the tool and not yet adjusted back to normal.

"This is the first evidence that tool use alters the body [map]," says Farné.

Farné says the same kind of brain "plasticity" might be involved in regaining control of a transplanted hand or a prosthetic limb when the original has been lost. The brain might also readily incorporate cyborg additions – a cyborg (part human, part robot) arm or other body part – into its body schema, says Farné, "and possibly new body parts differing in shape and/or number, for example four arms." Small implants such as pacemakers are inserted in the existing body so do not need to be accepted by the body schema, adds Farné, "but a pair of wings would – that would be tough!" Journal reference: current biology (DOI: 10.1016/j.cub.2009.05.009)

5. In the 1960s, Paul Bach-y-Rita invented a device that allowed blind people to read, perceive shadows, and distinguish between close and distant objects. This "machine was one of the first and boldest applications of neuroplasticity." The patient sat in an electrically stimulated chair that had a large camera behind it that scanned the area, sending electrical signals of the image to four hundred vibrating stimulators on the chair against the patient's skin. The six subjects of the experiment were eventually able to recognize a picture of the supermodel Twiggy. It must be emphasized that these people were congenitally blind and had previously not been able to see. Bach-y-Rita believed in sensory substitution; if one sense is damaged, your other senses can sometimes take over. He thought skin and its touch receptors could act as a retina (using one sense for another). In order for the brain to interpret tactile information and convert it into visual information, it has to learn something new and adapt to the new signals. The brain's capacity to adapt implied that it possessed plasticity. He thought, "We see with our brains, not with our eyes." [Doidge, Norman (2007). The Brian That Changes Itself: stories of personal Triumph from the frontiers of brain science. New York: Viking. ISBN 9780670038305]

6. David Hubel, Torsten Wiesel, and Michael Merzenich (a neuroscientist who has been one of the pioneers of brain plasticity for over three decades) performed an ambitious experiment. The experiment was based on an observation of what occurred in the brain when one peripheral nerve was cut and subsequently regenerated. The first two scientists micromapped the hand maps of monkey brains before and after cutting a peripheral nerve and sewing the ends together.

Afterwards, the hand map in the brain that was expected to be jumbled was nearly normal. This was a substantial breakthrough. Merzenich asserted that "if the brain map could normalize its structure in response to abnormal input, the prevailing view that we are born with a hardwired system had to be wrong. The brain had to be plastic.

7. The phantom limb phenomenon often occurs in people who have lost a limb through an accident or surgical amputation. Despite the fact that there is no longer a limb present, patients routinely report that they feel that their limb is still there. Sadly, they often find this "phantom limb" to be quite tender or painful. Note that this is different than pain at the edge of the remaining skin-the person actually feels that the limb is still there. This is situation eventually causes brain plasticity as the brain learns that there is no longer a limb present.

8. One ingenious experiment shows how our brain is rapidly able to adapt to strange stimuli. Wolfgang Kohler had subjects wear a special pair of glasses in which the lenses were prisms. These prisms inverted everything that was seen. What was up was down and down was up. The first few days of wearing these glasses were challenging as people had trouble getting around. After a few days however, they began to adjust and could do daily activities. Some could eventually ski or ride a bike! Wearing the glasses actually causes brain plasticity. How do we know? When the subjects stopped wearing the glasses, they did not immediately go back to normal but needed to adjust. Thus taking off the glasses causes brain plasticity, too!

9. Some patients claim that they can experience vivid voluntary movements in their phantom limb, presumably because reference signals from motor commands sent to the phantom limb are monitored in the cerebellum and parietal lobes. However, over time, the phantom limb becomes "frozen" or "paralyzed." Some patients experience excruciatingly painful involuntary clenching spasms in the phantom limb; they experience their nails digging into their phantom palm and are unable to open the hand voluntarily to relieve the pain.

Researchers placed a mid vertical sagittal mirror on the table in front of the patient, Figure (2.23). If the patient's paralyzed phantom limb was, say, on the left side of the mirror, he placed his right hand in an exact mirror symmetric location on the right side of the mirror. If he looked into the shiny right side of the mirror, changes in cortical topography revealed by magnetoencephalography. The reflection of his own right hand is optically superimposed on the felt location of his phantom limb so that he has the distinct visual illusion that the phantom limb had been resurrected. If he now made mirrorsymmetric movements while looking in the mirror, he received visual feedback that the phantom limb was obeying his command. Remarkably, 6 of 10 patients using this procedure claimed that they could now actually feel—not merely see—movements emerging in the phantom limb. This was often a source of considerable surprise and delight to the patient. Indeed, 4 patients were able to use the visual feedback provided to them by the mirror to "unclench" a painfully clenched phantom hand and this seemed to relieve the clenching spasm, as well as associated cramping pain (the burning and lancinating pains in the phantom limb remained unaffected by the mirror procedure, suggesting that the relief of the clenching was probably not confabulatory in origin). The elimination of the spasm is a robust effect that was confirmed in several patients. Patients reported the elimination of the associated pain but this requires confirmation with double-blind control subjects, given the notorious susceptibility of pain to placebo and suggestion. In one patient, repeated use with the mirror for 2 weeks resulted in a permanent and full disappearance of the phantom arm and elbow (and a "telescoping" of fingers into the stump) for the first time in 10 years. The associated pain in the elbow and wrist also vanished. This may be the first known instance of a successful amputation of a phantom limb!

Figure (2.23): Mirror box used to provide visual feedback. Patient views the reflection of his own hand in the mirror

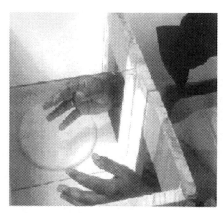

The patient places the good limb into one side of the box (in this case the right hand) and the amputated limb into the other side. Due to the mirror, the patient sees a reflection of the good hand where the missing limb would be. The patient thus receives artificial visual feedback that the "resurrected" limb is now

2.34.2 Brain Training and Fitness

The old idea was that brain structure was hard wired and permanent. After a young age, you had your fixed number of neurons and brain cells, with tens of thousands dying every day for the rest of your life. That's been discredited. Experiments show the creation of new neurons and brain cells even in the aged people.

The old idea was called brain decentralization and localization: particular parts of the brain perform particular things, like sight and hearing. If you lost those parts of the brain, you lost those functions permanently. The new idea is plasticity: functions can be carried out by different locations of the brain.

173

Norman Doidge is a psychiatrist who tells about brain training in his book, *The Brain that Changes Itself: Stories of Personal Triumph from the Frontiers of Brain Science* (New York: Viking, 2007).

"I met a scientist who enabled people who had been blind since birth to begin to see, another who enabled the deaf to hear; I spoke with people who had had strokes decades before and had been declared incurable, who were helped to recover with neuroplastic treatments; I met people whose learning disorders were cured and whose IQs were raised; I saw evidence that it is possible for eighty-year-olds to sharpen their memories to function the way they did when they were fifty-five. I saw people rewire their brains with their thoughts, to cure previously incurable obsessions and traumas."

With the advances of neuroscience, we have discovered many ways to master peak performance of our brains through 'reading' and mapping the brains of the greatest yogis, sports champions, and even creative geniuses. There are many people that have mastered peak performance through their long arduous brain training methods. Various brain technologies are revealing their secrets and we can now easily recreate those peak performance states in others.

2.34.3 Methods to Increase Intelligence

   a. Your brain needs a lot of oxygen and water to optimize its functions. It is estimated that our brain uses about one fifth of our entire oxygen consumption. Physical activity can increase the oxygen which contribute to intelligence, increase memory retention, improve some types of problem-solving ability, and help relieve stress.
   b. A proper diet is important for the proper function of the brain. For example, after large meals, our digestive processing uses up a good portion of the available oxygen in biochemical reactions and limit the oxygen percentage that the brain receives through the blood transport. A high fat diet can change our body clock and thereby disrupt a range of behavioural and physiological processes, including those controlled by the brain that switch on and off at certain times to keep the body's metabolism, energy and biochemical reactions required for life.
   c. Exercise the mind by playing games. It is believed that mental exercise, such as doing puzzles, problem solving, strategy games like chess, playing musical instruments, and solving IQ tests can help to maintain and increase the cognitive abilities.
   d. Read books on mathematics, science, art, philosophy, history, etc. Reading is an active mental process which improves your vocabulary, improves concentration and focus and improves memory and creativity.
   e. One of the amazing stories told by Norman Doidge is that physical changes in the brain can be brought about by thought alone. The usual idea is that the brain responds to external stimuli but that the mind cannot initiate change. This materialistic philosophy is challenged by studies by

Alvaro Pascual-Leone. In one experiment, he taught two groups of people some piano basics. One group practiced using actual pianos; the other group practiced only in their minds. After six hours of practice over three days, the two groups had similar changes in their brain maps and similar improvements in actual performance.

f. Amazingly, some cultures (South Eastern Asian) believe that you can strengthen your muscles by simply imagining that you're exercising them, or that you can reduce your weight by deeply thinking and imagining that your weight is really reduced.

3.34.4 Brain Trust Program

The brain, like the muscles, heart, and other organs, is made of flesh and blood and requires proper care to maintain its optimum state of wellbeing. The entire brain constitutes about 2 percent of the weight of the body yet consumes 20 percent of the energy, which means that it is ten times more metabolically active than the rest of the body.

The goal of the Brain Trust Program (BTP) is to improve the function of the brain and to prevent and forestall its decline with age.

The BTP offers a program that improves brain health. Not only can it improve brain function, but it can substantially reduce the risk of developing memory impairment and a host of other brain problems. The program contains a list of items recommended by notable neurosurgeons, psychiatrics, and dieticians. The list contains the following items:

1. Low fat diet – Too much fat produces too much ketones. The brain cannot use fatty acids for energy, and instead uses ketones produced from fatty acids by the liver. Simple ketones are generally not highly toxic to the brain, but too much ketones could have a negative effect on the brain function. A low fat diet is recommended.
2. Low carbohydrate diet – Sugar provides a quick boost in short-term energy but at a long-term cost. People whose diet contains a large amount of sugar, or foods that are rapidly turned into sugar, frequently have blood sugar problems in long term memory, and also much higher rates of memory loss and other degenerative brain disorders (including Alzheimer's). The connection is quite clear. A low carbohydrate diet is recommended.
3. Stress and anxiety – Elevated levels of cortisol (the stress hormone) could cause the death of brain cells that affect both memory and mood. High level of stress could kill the nerve cells of the hippocampus, the gatekeeper for memory, which is the first place where Alzheimer originates. Stress and anxiety should be avoided.

4. Menopause flashes and mood swings – These are brain driven and could cause disturbance in brain energy level and brain function. Proper nutrition and hormone regulation can smooth out such disturbances.
5. Diabetes or insulin resistance – increases the risk of developing memory loss with aging—even developing Alzheimer's disease. Patients with diabetes must control their sugar level and avoid elevated blood sugar.
6. Sleep deprivation – Lack of sleep (an average of sleeping about 4 hours a night) can disturb the metabolic systems that regulate blood sugar which could affect the brain function and the cognitive ability. You must sleep between 7 to 8 hours daily.
7. High blood pressure and high bad cholesterol – High blood pressure and high bad cholesterol levels could harden the blood vessel from the big aorta to the tiny coronary arteries that feed the heart, the carotid arteries and the vertebral arteries in the neck that supply the blood to the head and the brain. Hardening of the blood vessel is called arteriosclerosis. Good diet and medication can reduce the blood pressure to normal.
8. Smoking – Smoking is hazardous to the memory, heart, lungs, liver and body in general. One study, published in the scientific brain journal *Neurology* in 1999 showed that smoking doubles a person's risk for developing Alzheimer's disease and developing less severe forms of memory loss. Smoking is mostly toxic and should be avoided.
9. Alcohol use – Results of studies designed to test the association between alcoholism and confirmed dementia indicate that people having 1 drink a week lower their risk of memory decline by 35 percent over teetotalers; those who take 1 to 6 drinks a week reduce their risk by 54 percent. The protective effect begins to decline a bit to a 31 percent reduction in those who have 7 to 13 drinks a week, and the protection vanishes entirely thereafter. People who regularly drink 14 or more drinks a week have a 22 percent greater risk of developing dementia as they age. Avoid too much drinking of alcohol because it could deteriorate the memory.
10. Medical dugs – The brain is made mostly of delicate fats that are particularly susceptible to oxidization which could have serious threatening consequences due to the mutation or catabolism of lipids molecules. All fats and proteins are susceptible to oxidation by free radicals in the medical drugs. Protein and fat fragmentation are the products of oxidation.

2.34.5 Executive Brain

Using functional magnetic resonance imaging (fMRI), Nico Dosenbach and his colleagues (at the Washington University School of Medicine, St. Louis) have shown that the dorsal anterior cingulate cortex and operculum, Figure (2.24), were consistently activated when their 183 subjects began working on a cognitive task. The findings suggest that these areas are, effectively, the 'executives, which determine the order in which the brain regions and circuits contributing to a particular task are activated during the execution of that task.

Figure (2.24): Anterior cingulate cortex and operculum

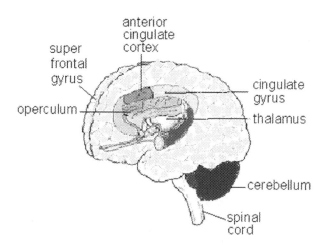

The executive brain is the parts of the brain that are responsible for abilities like vision, imagination, ethical and social behaviour, planning, cognitive flexibility, abstract thinking, rule acquisition, initiating appropriate actions and inhibiting inappropriate actions, and selecting relevant sensory information, Stuss, D. & Knight R.T. (Editors) (2002). *The Frontal Lobes*. New York: Oxford University Press. The executive brain is thought to be heavily involved in handling novel situations outside the domain of some of our 'automatic' psychological processes that could be explained by the reproduction of learned schemas or set behaviors.

The neural mechanisms by which the executive brain's functions are implemented are a topic of ongoing debate in the field of cognitive neuroscience and sending and transmitting its neurotransmitters. Traditionally, there has been a strong focus on the frontal lobes, but more recent brain research indicates that executive functions are far more distributed across the frontal cortex.

It has been suggested that there are five types of situations in which the routine activation of behavior would not be sufficient for optimal performance, Norman DA, Shallice T (2000). "(1980) Attention to action: Willed and automatic control of behaviour", Gazzaniga MS. *Cognitive neuroscience: a reader*. Oxford: Blackwell.

1. Those that involve planning or decision making.
2. Those that involve error correction or troubleshooting.
3. Situations where responses are not well-rehearsed or contain novel sequences of actions.
4. Dangerous or technically difficult situations.
5. Situations that require the overcoming of a strong habitual response or resisting temptation.

Effective executive function training is designed to improve the mental agility, foresight, ability to plan, maintain attention, and mental set shifting.

When executive function encounters with antisocial behaviour, it may be subjected to numerous psychiatric and developmental disorders, such as attention-deficit/hyperactivity disorder (ADHD), depression, schizophrenia, obsessive-compulsive disorder, autism, and Tourette's syndrome (Tourette's syndrome is characterized by repetitive, stereotyped, involuntary movements and vocalizations called tics).

Executive brain function appears to be dependent on experience and training. For example, stress and anxiety appear to slow down the creation of new neurons in the hippocampus, whereas using particular parts of the brain appears to make those parts grow. In one study by researchers in London, England, cab drivers (who have to pass a rigorous test demonstrating their knowledge of London streets) were found to have a larger posterior hippocampus than other men of the same age who did not have driving licenses. Moreover, the size of their posterior hippocampus corresponded closely to the number of years that they had been driving a cab. This finding suggests that relying heavily on spatial memory (and engaging regularly in navigation) leads to an increase in the size of the posterior hippocampus.

Each type of experience and training affects its own responsible part in the brain. For example, a study, by researchers in Germany, examined the sensorimotor cortices of violin players—specifically the part that is associated with the complex fingering movements involved in playing the violin. Results indicated that this part of the brain was increased in violin players compared to those who don't play stringed instruments.

Other studies were made on the sensory system. For example, people using their smell senses in the wine industry have thicker olfactory pathways and limbic lobes.

Neuroimaging using an fMRI (functional MRI) showed that the frontal cortex of the brain is mainly responsible for the brain is executive function. FMRI creates images that reflect increased blood flow in the frontal cortex that are more active than other areas in the brain - areas where neurons are "firing."

2.34.6 Perception Attention and the Four Theatre of the Brain

John Ratey, bestselling author and clinical professor of psychiatry at Harvard Medical School, explains the dynamic process and the interaction of the brain as a metaphor called the four "theaters" of process:

1. the act of perception
2. the filters of attention, consciousness, and cognition
3. the array of options employed by the brain--memory, emotion, language, movement -- to transform information into function
4. behavior and identity

Diagrammatically, the four theaters are represented by Figure (2.25).

Figure (2.25): The four theatres of the brain

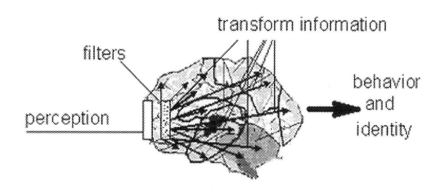

An interesting thing about Ratey's approach is that the brain is composed of theaters extended along a neurophysiological river of the mind, with each theater further downstream from immediate experience than the one before it.

By this, he achieves a number of things. For one, a ordinary person can begin to realize the relationship between brain centers, the way the environment interacts with our minds and bodies to help shape us.

Ratey insisted that that the brain is dynamic. While the theaters idea of the brain may not be universally accepted, it is indeed a handy way for an outsider to think of the brain and learn about it.

2.34.7 Peak Brain Performance

The first direct evidence that stress can shrink a crucial part of the human brain is being compiled by scientists using new, high-resolution magnetic resonance imaging (MRI) scans, according to a Stanford expert on stress and the brain.

In a review article in the Aug. 9 edition of the journal *Science*, biological sciences Professor Robert Sapolsky said that the work of several research groups shows links between long-term stressful life experiences, long-term exposure to hormones produced during stress, and shrinking of the part of the brain involved in some types of memory and learning, Stanford – Peak performance.

Sapolosky reported that a steroid hormone called glucocorticoid (glucose + cortex + steroid) can stimulate the fat breakdown in adipose tissue, and thus can cause rats' brain cells to shrink and the dendrites of the neurons to wither away. The glucocorticoid hormone is synthesized in the adrenal cortex, and virtually any type of physical or mental stress results in elevation of glucocorticoid concentrations in blood due to the enhanced secretion of CRH (corticotropin-releasing hormone) in the hypothalamus.

Cortisol secretion is suppressed by classical negative feedback loops. When blood concentrations rise above a certain threshold, cortisol inhibits CRH secretion from the hypothalamus, which turns off the ACTH (adrenocorticotropic hormone) secretion, which leads to a turning off of the cortisol secretion from the adrenal. The combination of positive and negative control on CRH secretion results in pulsating secretion of glucocorticoid. Typically, the pulse amplitude and frequency are highest in the morning and lowest at night. Therefore, enough sleep could reduce the level of the stress, thereby, reducing the shrinking of brain cells. Figure (2.26) shows the effect of the stress on the brain cells

Figure (2.26): Effect of glucocorticoid on the brain cells

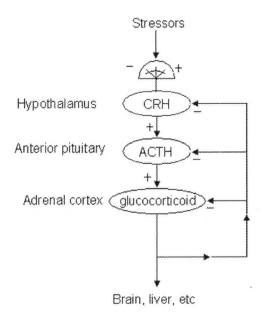

Because depression can raise the levels of glucocorticoids in the blood, Yvette Sheline and her colleagues at Washington University in St. Louis compared the hippocampi (the neurons of the hippocampus are rich in glucocorticoid receptors; this is the region where animal studies have shown that stress hormones can damage neurons) of people who had recovered from long-term, major depression with controls matched to them by age, education, gender and height. They found that the people with a history of depression had smaller hippocampi (averaging as much as 15 percent smaller in volume).

Cushing's syndrome (a tumor adenoma) in the pituitary gland produces large amounts of ACTH, which in turn elevates glucocorticoids in the adrenal glands to produce large quantities of glucocorticoids. When Monica Starkman of the

University of Michigan at Ann Arbor scanned the brains of people with this disorder, she found that the hippocampus had atrophied in the same areas where glucocorticoids were being over-produced. In a neighboring region of the brain that has few glucocorticoid receptors, there was no atrophy.

In addition to the above adverse effect, prolonged production of large quantity of glucocorticoids may cause a number of adverse effects. These include the suppression of the immune system (which makes the person more susceptible to infections), osteoporosis, shifts in the body's fluid balance, skin changes, changes in brain chemistry, and altered behavior. However, glucocorticoids are used to stop the inflammation process and to treat severe acute respiratory syndrome (SARS), rheumatoid arthritis, and osteoporosis if the regulated quantity is given to the patient.

Factors generally reducing cortisol (glucocorticoid) levels are:

- Magnesium supplementation decreases serum cortisol levels after aerobic exercise, but not in resistance training.
- Omega 3 fatty acids, in a dose dependent manner (but not significantly), can lower cortisol release influenced by mental stress by suppressing the synthesis of interleukin-1-1 and 6 and enhance the synthesis of interleukin-2, where the former promotes higher CRH release. Omega 6 fatty acids, on the other hand, act inversely on interleukin synthesis.
- Music therapy can reduce cortisol levels in certain situations.
- Message therapy can reduce cortisol.
- Sexual intercourse can reduce cortisol levels.
- Laughing and the experience of humour can lower cortisol levels.
- One study by a Japanese cosmetics company has asserted that makeup reduces cortisol levels in a mental stress situation.
- Soy derived Phosphatidylserine interacts with cortisol but the right dosage is still unclear.
- Vitamin C may slightly blunt cortisol release in response to a mental stressor.
- Black tea may speed up recovery from a high cortisol condition.

Peak Brain Performance can also be optimized by specific training programs and brain teasers that test your mental acuity such as:

1. Cognitive training: By taking a series of tests (IQ and EQ tests) to determine your areas of strength and weakness in your thinking skills. Then, try to strengthen those areas where your scores are low by practicing and solving problems.
2. Cognitive sporting strategies: To control and utilize an athlete's mental and mind game. Cardiovascular exercise improves blood flow to the brain which improves brain functioning.

3. Reading: Reading books, magazines, and newspapers stimulates the imagination and mental acuity.
4. Enough sleep: A lack of sleep can interfere with your ability to assimilate and digest new information.
5. Lay off the alcohol and cigarettes: Alcohol is of course a depressant and slows down mental functions. Nicotine reaches the brain within 10 seconds after smoke is inhaled. It has been found in every part of the body and in breast milk. Smoking is associated with higher levels of chronic inflammation, another damaging process that may result in oxidative stress.
6. Expand your vocabulary and speak a new tongue: Learning new words and to speak a new language are powerful ways to give the mind more elasticity and expanse.
7. Profound relaxation: Music, yoga and meditation can access an exquisite state of mental calm and expanded awareness with ascension and commune with higher aspects of self.
8. Gaming the brain: Games are proven to improve brain function while targeting the five major cognitive brain functions - memory, concentration, language, executive functions (logic and reasoning), and visual-spatial skills. Research has proven that the brain begins to slow down as early as age 25 but with regular brain exercise, it can create new neural connections through new dendrites and pathways at any age.
9. Balanced electrolytes: Lack of sodium and potassium electrolytes can reduce critical brain function. Balance of electrolytes is also important with potassium being needed in larger quantities.
10. Nutritional food: Low G.I Carbohydrates and high fat meals could create fluctuations in blood sugar levels and may create brain cell damage.
11. Multi vitamin and minerals: A good multi vitamin and mineral supplement can act as sound boost for brain function.

Chapter 3

Brain Diseases

3. Brain Diseases

The brain is the control center of the body. It controls thoughts, memory, speech and movement. It regulates the function of many organs. When the brain is healthy, it functions quickly and automatically. However, when problems occur, the results can be devastating. Some of the major types of disorders include stroke, epilepsy, myelin disorders, toxic encephalopathies, genetic diseases such as Alzheimer (causes of which is not known) affecting the brain, alcohol and brain metabolic diseases such as Gaucher's diseases, trauma such as spinal cord, viral and nonviral encephalitides, cerebrovascular diseases such as strokes and vascular dementia, diseases affecting neurotransmitters, infectious diseases such as AIDS dementia, Alzheimer's disease, and aging.

Inflammation in the brain can lead to problems such as vision loss, weakness and paralysis. Loss of brain cells, which happens if you suffer a stroke, can affect your ability to think clearly. Brain tumors can also press on nerves and affect brain function.

Medications can be prescribed to slow the progress of some brain diseases, but success is often coupled with unwanted side effects, particularly in the elderly patient. Medication can also be used in conjunction with psychiatric or behavioral therapies. In some cases, treatments such as surgery or physical therapy can correct the source of the problem or improve symptoms.

Neurology is the medical specialty encompassing diseases, conditions, and infections of the nervous system, which includes the brain, spinal cord, and peripheral nerves.

Neurologists have found more than 135 brain diseases including, autism, chronic pain, schizophrenia and dementia that are linked to defects in proteins in the junctions between nerve cells, (the synapses).

Humans have around a billion brain cells and these are connected by synapses, which play a pivotal role because they create circuits that allow the brain to learn and remember things. Neurologists found around 1,500 proteins in human synapses, each of which is encoded by a gene. They then managed to link genetic defects in some of these with key diseases such as autism, bipolar disorder, depression and schizophrenia.

Diseases are caused by the gene mutation that disrupts the protein formula. It is found that defects in the genes that encode these human synapse proteins are really a major cause of diseases. Some of the brain diseases are discussed below:

3.1 Central Auditory Disease

Central auditory processing disorders have been found in many cases when patients with Alzheimer's disease are tested. They may present as an early manifestation of Alzheimer's disease, preceding the disease by a minimum of 5 and a maximum of 10 years. A condition in which there is an inability to differentiate, recognize or understand sounds while both the hearing and intelligence are normal. The problem is "central" as regards to the auditory pathways. It is a disease of the auditory pathways from the bulbar cochlear nuclei to the auditory cortex in the temporal lobe, Figure (3.1). Structures involved in such a disorder may include the medial and lateral lemnisci, inferior colliculus, and the medial geniculate nucleus.

Figure (3.1): Connection of auditory path to the brain

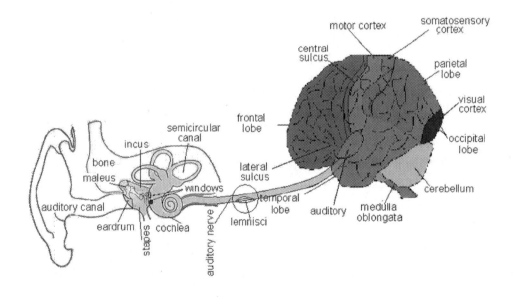

3.2 Brain Neoplasms

Brain neuroplasms are tumors or tissues containing a growth in the brain. Neoplasms of the intracranial components of the central nervous system (CNS), include the cerebral hemispheres, basal ganglia, hypothalamus, thalamus, brain stem, and cerebellum. Brain neoplasms are subdivided into primary (originating from brain tissue) and secondary (i.e., metastatic, i.e., a malignant tumor that has developed in the brain as a result of the spread of cancer cells from the original tumor, for example, from lung cancer) forms. Primary neoplasms are subdivided into benign and malignant forms. In general, brain tumors may also be classified by age of onset, histologic type, or presenting location in the brain.

Some of the causes of Brain neoplasm are:

a. Neurofibromatosis

Neurofibromatosis is a complicated genetic disease that can affect both men and women in all races and ethnic groups. The neurofibromatoses are a set of genetic disorders that cause tumors to grow on and along various types of nerves. The tumors can grow anywhere on or in the body. The disease can also affect non-nervous tissues like bones and skin and lead to developmental abnormalities such as learning disorders, according to the National Neurofibromatosis Foundation (NNFF). Scientists have classified Neurofibromatosis (NF) into two distinct types:

- NF1: This is the more common type of the neurofibromatosis, occurring in 1 in 4,000 births. It is sometimes called Von Recklinghausen NF or Peripheral NF. In most cases, symptoms of NF1 are mild, and patients live normal and productive lives. In some cases, however, NF1 can be severely debilitating.
- NF2: This form of the neurofibromatosis is much rarer, occurring in 1 in 40,000 births. It is also know as Bilateral Acoustic NF (BAN). It is characterized by multiple tumors on the cranial and spinal nerves and by other lesions of the brain and spinal cord. People with NF2 are also at high risk for developing brain tumors and in severe cases, the disease can be life threatening.

Both forms are genetic disorders that may be inherited from a parent, but according to the National Institute of Neurological Disorders and Stroke (NINDS), 30 to 50 percent of the cases are the result of a gene mutation in an individual rather than it being inherited.

b. Tuberous Sclerosis

Tuberous sclerosis (TSC) is a rare genetic disease that causes benign tumors to grow in the brain and on other vital organs such as the kidneys, heart, eyes, lungs, and skin. It commonly affects the central nervous system. In addition to the benign tumors that frequently occur in TSC, other common symptoms include seizures, mental retardation, behavior problems, and skin abnormalities. TSC may be present at birth, but signs of the disorder can be subtle and full symptoms may take some time to develop. Three types of brain tumors are associated with TSC: cortical tumors, which generally form on the surface of the brain; subependymal nodules, which form in the walls of the ventricles (the fluid-filled cavities of the brain); and giant-cell astrocytomas, a type of tumor that can block the flow of fluids within the brain.

185

c. Tuberculosis

Tuberculosis (It was first isolated in 1882 by a German physician named Robert Koch who received the Nobel Prize for this discovery) is a contagious disease caused by a bacterial infection of the lungs, which can also spread to other parts of the body, such as the brain, kidneys, and bones. Tuberculosis, also known as TB, is caused by the bacterium Mycobacterium tuberculosis. Tuberculosis is contagious and spreads to others when an infected person coughs or sneezes. This shoots droplets contaminated with Mycobacterium tuberculosis bacteria into the air where they can be breathed in by others.

Tuberculosis that infects the tissues covering the brain (tuberculous meningitis) is life threatening. In the United States and other developed countries, tuberculous meningitis most commonly occurs among older people or people with a weakened immune system. In developing countries, tuberculous meningitis is most common among children from birth to age 5. Symptoms include fever, constant headache, neck stiffness, nausea, and drowsiness that can lead to coma. Tuberculosis may also infect the brain itself, forming a mass called a tuberculoma. The tuberculoma may cause symptoms such as headaches, seizures, or muscle weakness. A tuberculoma in the brain may need to be surgically removed.

d. Syphilis

Syphilis is a type of many different sexually transmitted diseases. Syphilis is the result of a bacterial infection of the genital tract by the bacterium treponema pallidum. Syphilis is passed from one person to another during direct sexual contact with a syphilis lesion that involves vaginal, oral, or anal sex. The highly infectious disease may also be passed, this happens through blood transfusions or from mother to fetus in the womb and result in stillbirth or serious birth defects. Syphilis can cause irreversible damage to the brain, nerves, and body tissues.

The symptoms of syphilis can mimic many diseases. Sir William Osler stated, "The physician who knows syphilis knows medicine."

The symptoms of syphilis depend on the stage of the disease. Many people do not have symptoms. In general, painless sores and swollen lymph nodes are possible symptoms of primary syphilis (the first stage). Those with secondary syphilis (the second stage) may also have fever, fatigue, rash, aches and pains, and loss of appetite, among other symptoms. Tertiary syphilis (the third and last stage) causes heart, brain, and nervous system problems.

e. Neurofibromatosis, type 4, of Riccardi

Neurofibromatosis, type 4, of Riccardi is a rare genetic disorder characterized by areas of increased and decreased skin pigmentation and the development of many non-cancerous nerve and skin tumors some of which may eventually become malignant.

A neurofibroma is a mass that contains multiple cell types-Schwann cells, perineurial cells, nonspecific fibroblasts, endoneurial fibroblasts, endothelial cells, mast cells, sometimes adipocytes, smooth muscle cells, melanocytes, and hair cells-and as well large amounts of intercellular material, including collagen and amorphous ground substances. Most physicians would refer to a neurofibroma as a tumor, however, in some cases it is not.

3.3 Hypothalamic Diseases

The hypothalamus regulates homeostasis. It has regulatory areas for thirst, hunger, body temperature, water balance, sexual development, blood pressure, and many somatic functions links the nervous system to the endocrine system. The hypothalamus also regulates the functions of the pituitary gland by directing the pituitary to stop or start production of its hormones. The hypothalamus is involved in many functions of the autonomic nervous system, as it receives information from nearly all parts of the nervous system. As such, it is considered the link between the nervous system and the endocrine system.

Hypothalamic diseases could affect one or more of the following:

- Heart rate and blood pressure
- Body temperature
- Fluid and electrolyte balance, including thirst
- Appetite and body weight
- Glandular secretions of the stomach and intestines
- Production of substances that influence the pituitary gland to release hormones
- Sleep cycles
- Autonomic nervous system

3.4 Akinetic Mutism

Akinetic mutism is a state in which a person is unspeaking (mute) and unmoving (akinetic). Akinetic mutism is often due to damage to the frontal lobes of the brain. It is the result of many possible causes such as:
- Frontal lobe injury – in the bilateral medial frontal lobe which contains cingulate gyrus
- Basal ganglia impairment – ventral striatum, globus palldus

- Fornex/limbic system malfunction
- Thalamus malfunction
- White matter diffusion (leukoencephalopathy)
- Ablation of cingulate gyrus (destruction and ablation of the cingulate gyrus has been used in the treatment of psychosis).

The above causes could also result in:

- Slow responses - lack of initiative & spontaneity (abulia)
- Psychiatric disease - immobile & mute (catatonia)
- sad and/or irritable mood (depression)
- inability to move or talk due to quadriparesis & bulbar paralysis (locked-in state)
- loss of cortical function (negative state/passivity)
- indifference to painful stimuli (congenital universal insensitivity to pain)

3.5 Epilepsy

Epilepsy is a brain disorder that causes people to have recurring seizures. These seizures are transient signs and/or symptoms of abnormal, excessive or synchronous neuronal activity in the brain. People may have strange sensations and emotions or behave strangely. They may have violent muscle spasms or lose consciousness.

People have seizures when the electrical signals in the brain misfire: they either fire when they shouldn't or don't fire when they should. Seizures could also occur when the brain's normal electrical transmission is disturbed by these overactive electrical pulses, causing a temporary disturbance problem between nerve cells.

The result is a sudden, brief, uncontrolled burst of abnormal electrical activity in the brain. Seizures are the physical effects of such unusual bursts of electrical energy in the brain, Figure (3.2).

Figure (3.2): Electrical pulses during normal and seizure modes

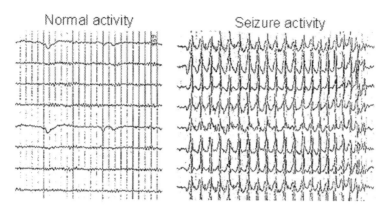

http://www.epilepsymatters.com/english/faqexplaining.html

Epilepsy has many possible causes, including illness, brain injury and abnormal brain development. In many cases, the cause is unknown. Determining the specific cause for any one person's epilepsy is usually difficult. In about 60% of all cases, no specific cause is found, much to the frustration of the epilepsy patients involved. Epilepsy of an unknown origin is called *idiopathic* epilepsy. In many cases it is presumed to be genetic.

3.6 Brain Edema

When a patient has cerebral edema, there is way more fluid in the skull than there should be. This causes the brain to swell, which has a number of consequences. As the brain swells, it can compromise its own blood flow. Decreased flow of blood to the brain can cause brain damage or death. The increased pressure in the skull may also force the brain to move around in the skull, which it is not designed to do.

Edema is seen in response to trauma, tumors, focal inflammation, late stages of cerebral ischemia and hypertensive encephalopathy.

The following categories can contribute to the brain edema:

- Hydrostatic cerebral edema is a life-threatening condition that develops as a result of an inflammatory reaction. This is due to the direct transmission of pressure to the cerebral capillary with transudation of fluid into the extracellular fluid from the capillaries.
- Cerebral edema from brain cancer is also a life-threatening condition. Most frequently, this is the consequence of cerebral trauma, massive cerebral infarction, hemorrhages, abscess, tumor, allergy, sepsis, hypoxia, and other toxic or metabolic factors.

- Osmotic cerebral edema is due to the pathophysiological variations of circulating osmolarity (including acute hyponatremia and hypernatremia) which can change the osmolarity of cerebral extra cellular fluids (ECFs).
- Cytotoxic cerebral edema is due to the derangement in cellular metabolism resulting in inadequate functioning of the sodium and potassium pump in the glial cell membrane.
- High altitude cerebral edema
- Interstitial cerebral edema

3.7 Amblyopia

Amblyopia, otherwise known as lazy eye (or "Lazy brain" is a more accurate term to describe amblyopia) is a disorder of the visual system that is characterized by poor or indistinct vision in an eye that is otherwise physically normal, or out of proportion to associated structural abnormalities. It has been estimated to affect 1–5% of the population. Amblyopia is the medical term used when the vision in one of the eyes is reduced because the eye and the brain are not working together properly. The eye itself looks normal, but it is not being used normally because the brain is favouring the other eye.

Amblyopia is caused by either no transmission or poor transmission of the visual stimulation through the optic nerve to the brain for a sustained period of dysfunction or during early childhood. This results in poor or dim vision.

The preferred eye has normal vision, but because the brain ignores the other eye, a person's vision ability does not develop normally. Between ages 5 and 10, the brain stops growing and the condition becomes permanent.

Strabismus (squinting, squint-eyed) is the most common cause of amblyopia, and there is often a genetic factor of this condition.

Other causes include:
- Astigmatism in both eyes – astigmatism is a vision disorder in which the eye focuses light on the retina at two points instead of just one, or the cornea may have areas that are flatter or steeper than others (irregular curvature), resulting in distorted vision.
- Childhood cataracts (clouding in the lens)
- Farsightedness
- Nearsightedness

People who have severe amblyopia may experience associated visual disorders, most notably poor depth perception. Amblyopes may suffer from poor spatial acuity, low sensitivity to contrast and some "higher-level" deficits to vision such as reduced sensitivity to motion.

3.8 Basal Ganglia Diseases

The main components of the basal ganglia are the striatum, pallidum, substantia nigra, and subthalamic nucleus, Figure (3.3). The largest component, the striatum, receives input from many brain areas but sends output only to other components of the basal ganglia. The pallidum receives its most important input from the striatum (either directly or indirectly), and sends inhibitory output to a number of motor-related areas, including the part of the thalamus that projects to the motor-related areas of the cortex. The basal ganglia control all signals to the thalamus, which in turn, control the signal to the cerebral. The cerebral, which controls the motor cortex, sends the signal to the putamen which controls the globus pallidus (medial part).

Figure (3.3): Basal ganglia in the brain

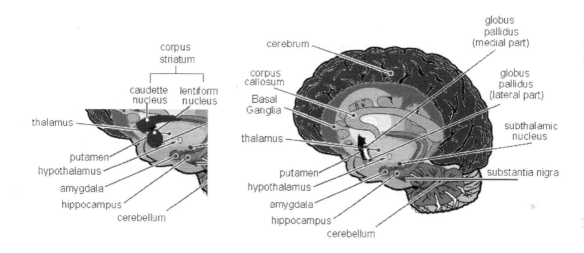

The same signal goes between the medial part and the lateral part of the globus pallidus to control the subthalamic nucleus. Figure (3.4) shows the pathway of the electrical signal.

Figure (3.4): Pathway of electrical signals in the brain

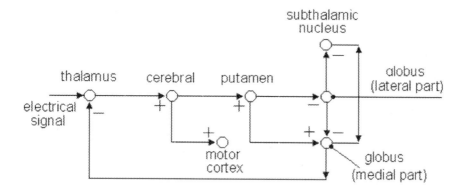

Diseases affecting the basal ganglia appear to be that the balance between the electrical signals is disturbed: the result is either involuntary movements or impairments to motion, which is controlled by the motor cortex. The impaired motions include lack of movement (akinesia), slowness of movement (bradykinesia), and the shuffling gait of Parkinson's disease.

Diseases of the basal ganglia include:

- Parkinson's disease which is believed to be related to low levels of dopamine in certain parts of the brain. It is related to damage to the dopaminergic pathway from the substantia nigra to the striatum. The result is an increased output of the basal ganglia to the thalamus. The major symptoms are tremor, rigidity, and akinesia.

- Huntington's disease which is caused by a loss of specific striatal neurons. The result is a decreased output of the basal ganglia to the thalamus (as opposed to the increased output due to Parkinson's). As one might expect from their opposing neural behavior, Huntington's disease results in hyperkineticity (the opposite of the hypokineticity of Parkinson's disease). The symptoms of HD include emotional turmoil, mental loss and/or physical deterioration. The disease leads to complete incapacitation and, eventually, death.

- Ballism which is tied to damage to the subthalamic nucleus. It causes violent flinging movements of the limbs, sometimes affecting only one side of the body (hemiballismus).

- Tardive dyskinesia which is related to changes in the dopaminergic receptors, causing hypersensitivity to dopamine. It causes abnormal involuntary movements, particularly of the tongue, lips, face, trunk, and extremities.

3.9 The Cerebellar Diseases

Cerebellum has two major motor roles:
1- Comparing the actual motor output to the intended movement.
2- Adjustment of movement as necessary.

The main clinical features of cerebellar disorders include incoordination, imbalance, and troubles with stabilizing eye movements.
There are two distinguishable cerebellar syndromes -- midline and hemispheric.

- Midline cerebellar syndromes: They are characterized by imbalance. Patients with midline lesions were unable to discern a global motion vector in a local stochastic motion display. They are unable to stand in Romberg (Romberg's test is a test used by doctors in a neurological examination, and also as a test for drunken driving) with eyes open or closed, and are unable to well perform the tandem gait. Severe midline disturbance causes "trunkal ataxia" a syndrome where a person is unable to stand, walk or sit on their bed without steadying themselves. Some persons have "titubation" or a bobbing motion of the head or trunk. Midline cerebellar disturbances also often affect eye movements (impaired smooth pursuit, dysmetria of saccades, and nystagmus).

- Hemispheric cerebellar syndromes: Cerebellar hemisphere lesions can produces classic limb ataxia (intention tremor, past pointing and mild hypotonia). Limb rebound can be demonstrated by gently pushing down on outstretched arms and then suddenly releasing, causing the arm on the affected side suddenly to fly upwards. Patients with HCS can have gait ataxia, truncal ataxia, limb ataxia, cerebellar dysarthria ((jerky and explosive speech with separated syllables), rapid alternating movements, and tremor.

3.10 Encephalitis

Encephalitis is an acute infection and inflammation of the brain itself. This is in contrast to meningitis, which is an inflammation of the layers covering the brain.

Encephalitis is generally a viral illness. Inflammation is usually caused by infection or an inappropriate auto-immune response to infection. Viruses like those responsible for causing cold sores, mumps, measles, and chickenpox can also cause encephalitis. A certain family of viruses, the Arboviruses is spread by insects like mosquitoes and ticks. The equine, West Nile, Japanese, La Crosse, and St. Louis encephalitis viruses are all mosquito-borne. Although viruses are the most common source of infection, bacteria, fungi, and parasites can also be responsible.

Encephalitis occurs in two forms — a primary form and a secondary form. Primary encephalitis involves direct viral infection of your brain and spinal cord. In secondary encephalitis, a viral infection first occurs elsewhere in your body and then travels to your brain.

Most people infected with viral encephalitis have only mild, often flu-like symptoms, and the illness usually doesn't last long. In some cases, people might not have any symptoms. Possible symptoms include headache, irritability, lethargy, fever, and joint pain.

More serious infections can cause confusion and hallucinations, personality changes, double vision, seizures, muscle weakness, loss of sensation or paralysis in certain areas, tremors, rash, loss of consciousness, and bulging in the soft spots (fontanels) of the skull in infants.

3.11 Dementia

Dementia means, in Latin, without mind. It is not merely a problem of memory.

Dementia is a gradual decline of mental ability that affects your intellectual and social skills to the point where daily life becomes difficult. Dementia can affect your memory and your decision-making ability, can impair your judgment and make you feel disoriented, and it may also affect your personality.

The Canadian Medical Association Journal has reported that driving with dementia could lead to severe injury or even death to self and others. Doctors should advise appropriate testing on when to quit driving, Neuropathology Group. Medical Research Council Cognitive Function and Aging Study (2001). "Pathological correlates of late-onset dementia in a multicentre, community-based population in England and Wales. Neuropathology Group of the Medical Research Council Cognitive Function and Ageing Study (MRC CFAS)". *Lancet* 357 (9251): 169–75.

Dementia may be caused by a number of factors, such as:

- Alcoholism
- Brain injury
- Drug abuse
- Side effects to certain medications like immunosuppressant
- Thyroid function abnormalities
- Vitamin B12 deficiency
- High blood pressure
- Brain tumor
- High cholesterol which could cause brain stroke
- Diabetes
- Smoking

Dementia could be caused by other factors including genetic factors, exposure to toxins, abnormal protein production, viruses, and difficulties in blood flow to the brain. Aging and heredity (genetic factors) are considered the greatest factors involved in the development of Alzheimer's disease.

Some causes of dementia can be treated. However, once brain cells have been destroyed, they cannot be replaced. Treatment may slow or stop the loss of more brain cells.

3.12 Klüver-Bucy Syndrome

Klüver-Bucy syndrome is a behavioral disorder that associated with bilateral lesions in the anterior temporal horn or amygdala. It causes individuals to put objects in their mouths and engage in inappropriate sexual behavior. Other symptoms may include visual agnosia (inability to visually recognize objects), loss of normal fear and anger responses, memory loss, distractibility, seizures, and dementia. The disorder may be associated with herpes encephalitis and trauma, which can result in brain damage. The amygdala has been a particularly implicated brain region in the pathogenesis of this syndrome. The syndrome is named after Heinrich Klüver and Paul Bucy.

Heinrich Klüver and Paul Bucy first described the syndrome in 1937 following experimental work where they removed rhesus monkeys' temporal lobes (the temporal lobes, one on each side of the head, just above the ears, are the sites of one of the most common forms of epilepsy. Complex partial seizures with automatisms (unconscious actions), such as lip smacking or rubbing the hands together, are the most common seizures in temporal lobe epilepsy.) , Kluver H, Bucy PC. Psychic blindness and other symptoms following bilateral temporal lobectomy in rhesus monkeys; *Am J Physiol* 1937; 119:352-3. They found that the monkeys developed:

1. Visual agnosia - they could see, but were unable to recognise familiar objects or their use.
2. Oral tendencies - they would examine their surroundings with their mouths instead of their eyes. This is called hyperphagia which is a strong compulsion to place objects in the mouths, probably to gain oral stimulation and to explore the object to counteract the visual agnosia, rather than due to hunger. Nevertheless, there is bulimia (an eating disorder) and there will be marked weight gain unless diet is restricted. Actions may include socially inappropriate licking or touching.
3. Hypermetamorphosis - a desire to explore everything and to the tendency to react to every visual stimulus.
4. Emotional changes - emotion was dulled and facial movements and vocalizations were far less expressive. They lost fear where it would normally occur. Even after being attacked by a snake, they would casually approach it again. This was called "placidity".
5. Hypersexualism - a dramatic increase in overt sexual behaviour including masturbation, homosexual and heterosexual acts. They may even attempt copulation with inanimate objects. They have an overactive libido and an obsession with sex

3.13 Cerebrovascular Disorders

Cerebrovascular (cerebral refers to the brain and vascular refers to blood vessels) disease is a group of brain dysfunctions related to disease of the blood vessels supplying the brain. Hypertension is the most important cause; it damages the blood vessel lining and endothelium, exposing the underlying collagen where platelets aggregate to initiate a repairing process which is not always complete and perfect. Strokes can result from blood vessel conditions such as aneurysms (blood vessel weakness which can result in a rupture) and thrombosis (blood vessel blockage). Sustained hypertension permanently changes the architecture of the blood vessels making them narrow, stiff, deformed, uneven and more vulnerable to fluctuations in blood pressure. Atherosclerosis is a common cerebrovascular condition where fatty deposits are laid down inside arteries causing them to become increasingly narrowed.

Causes of cerebrovascular disorders could be one or more of the followings:

- Cardiovascular disease
- Elevated cholesterol or elevated hematocrit (percentage of red blood cells to the total blood volume)
- Obesity
- Diabetes
- Oral contraceptive use
- Hypertension

Patients with cerebrovascular can be exposed to:

- Hemorrhagic stroke –Hemorrhagic strokes include bleeding within the brain (intracerebral hemorrhage) and bleeding between the inner and outer layers of the tissue covering the brain (subarachnoid hemorrhage). There are two main types of hemorrhagic strokes: intracerebral hemorrhage and subarachnoid hemorrhage. Other disorders that involve bleeding inside the skull include epidural and subdural hematomas, which are usually caused by a head injury. These disorders cause different symptoms and are not considered strokes

- Ischemic stroke – An ischemic stroke is the death of an area of brain tissue (cerebral infarction) resulting from a decreased supply of blood and oxygen to the brain due to blockage of an artery. An ischemic stroke usually results when an artery to the brain is blocked, often by a blood clot or a fatty deposit due to atherosclerosis. Symptoms occur suddenly and may include muscle weakness, paralysis, lost or abnormal sensation on one side of the body, difficulty speaking, confusion, problems with vision, dizziness, and loss of balance and coordination.

    There are four reasons why this might happen:

    1. Thrombosis (obstruction of a blood vessel by a blood clot forming locally)
    2. Embolism (obstruction due to an embolus from elsewhere in the body)
    3. Systemic hypoperfusion (general decrease in blood supply)
    4. Venous thrombosis

3.14 Encephalomalacia

Encephalomalacia is due to abnormal softness of the cerebral parenchyma (the tissue characteristic of an organ, as distinguished from associated connective or supporting tissues) often due to ischemia (restriction in blood supply), infarction (tissue death due to a local lack of oxygen), infection, or craniocerebral trauma.

Encephalomalacia is also a term that means scar tissue in the brain. It results from some injury that damages the brain such as a stroke or head injury.

The term encephalomalacia is also used at times to refer more generally to degenerative conditions affecting the matter of the brain. If the condition affects the white matter of the brain, it is called leukoencephalomalacia. If it affects the gray matter, it is known as polioencephalomalacia.

A patient with encephalomalacia will experience a loss of brain function. The symptoms of encephalomalacia can be one or more of the following:

- Somnolence (extreme drowsiness)
- Blindness
- Ataxia (wobbliness and lack of coordination)
- Sleep walking
- Head pressing
- Circling
- Terminal coma
- Memory loss and mood swings

With encephalomalacia, unfortunately, once injury to the brain has occured, neurons (brain cells) are lost and are not regrown. For now there are no current treatments available to treat this type of brain injury and replace injured brain cells.

3.15 Hydrocephalus

Hydrocephalus, known as "water on the brain", is a medical condition in which there is an abnormal accumulation of cerebrospinal fluid (clear fluid that surrounds the brain and spinal cord) in the ventricles, or cavities, of the brain. This may cause increased intracranial pressure inside the skull and progressive enlargement of the head, convulsion, and mental disability. Hydrocephalus can also cause death.

The ventricular system is made up of four ventricles connected by narrow passages. Normally, cerebrospinal fluid (CFS) flows through the ventricles, exits into cisterns (closed spaces that serve as reservoirs) at the base of the brain, bathes the surfaces of the brain and spinal cord, increases the pressure on the brain and then reabsorbs into the bloodstream.

The cerebrospinal fluid has four important life-sustaining functions:

1. Keeping the brain tissue buoyant, acting as a cushion or "shock absorber",
2. Acting as the vehicle for delivering nutrients to the brain and removing waste,
3. Flowing between the cranium and spine compensates for changes in intracranial blood volume (the amount of blood within the brain), and
4. Keeping the temperature stable in the cranium.

Hydrocephalus may be congenital or acquired. Congenital hydrocephalus is present a birth and may be caused by either events or influences that occur during fetal development, or genetic abnormalities. An obstruction within the brain is the most frequent cause of congenital hydrocephalus. Acquired hydrocephalus develops at the time of birth or at some point afterward. Acquired hydrocephalus may result from other birth defects such as spina befida, conditions related to prematurity such as intraventricular hemorrhage (bleeding within the brain), infections such as meningitis, or other causes such as head trauma, tumors, and cysts. This type of hydrocephalus can affect individuals of all ages and may be caused by injury or disease.

There are three different types of hydrocephalus: communicating hydrocephalus, noncommunicating hydrocephalus, and normal pressure hydrocephalus. Communicating hydrocephalus is the most common type and occurs when the flow of CSF is blocked after it exits the ventricles. This form is called communicating because the CSF can still flow between the ventricles, which remain open. This blockage prevents the movement of CSF to its drainage sites in the subarachnoid space just inside the skull. Non-communicating hydrocephalus - also called "obstructive" hydrocephalus - occurs when the flow of CSF is blocked along one or more of the narrow passages connecting the ventricles. One of the most common causes of hydrocephalus is "aqueductal stenosis." In this case, hydrocephalus results from a narrowing of the aqueduct of Sylvius, a small passage between the third and fourth ventricles in the middle of the brain. Normal pressure hydrocephalus is marked by ventricle enlargement without an apparent increase in CSF pressure. Normal pressure hydrocephalus can happen to people at any age, but it is most common among the elderly. It may result from a subarachnoid hemorage, head trauma, infection, tumor, or complications of surgery. However, many people develop normal pressure hydrocephalus even when none of these factors are present for reasons that are unknown. There is another type of hydrocephalus which is termed "ex-vacuo" and occurs when stroke or traumatic injury cause damage to the brain. In these cases, brain tissue may actually shrink.

198

3.16 Idiopathic Intracranial Hypertension

Idiopathic intracranial hypertension (IIH), sometimes called by the older names benign intracranial hypertension (BIH) or pseudotumor cerebri (PTC), is a neurological disorder that is characterized by increased intracranial pressure (pressure around the brain) in the absence of a tumor or other diseases. The main symptoms are headache, nausea and vomiting, as well as pulsatile tinnitus (buzzing and ringing in the ears synchronous with the pulse), double vision and other visual symptoms. If untreated, it may lead to swelling of the optic disc in the eye, which can progress to vision loss, Binder DK, Horton JC, Lawton MT, McDermott MW (March 2004). "Idiopathic intracranial hypertension". Neurosurgery 54 (3): 538–51; discussion 551–2.

Idiopathic intracranial hypertension typically occurs in women of childbearing age. Incidence is 1/100,000 in normal-weight women but 20/100,000 in obese women. Intracranial pressure is elevated (> 250 mm H_2O); the cause is unknown but probably involves obstruction of cerebral venous outflow.

Almost all patients have a daily or near daily generalized headache of fluctuating intensity, at times with nausea. They may also have transient obscuration of vision or diplopia (Diplopia is the medical term for double vision). In multiple sclerosis, it is usually caused by lesions in the brainstem where the cranial nerves serving the eye muscles arise. Vision loss begins peripherally and may not be noticed by patients until late in the course.

Bilateral papilledema (swelling of the head of the optic nerve) is common; a few patients have unilateral or no papilledema. In some asymptomatic patients, papilledema is discovered during routine ophthalmoscopic examination. Neurologic examination may detect partial 6th cranial nerve palsy but is otherwise unremarkable.

3.17 Intracranial Hypotension

Intracranial hypotension is a condition in which there is negative pressure within the brain cavity. There are several possible causes:

1. A leak of CSF following a lumbar puncture (spinal tap)
2. A defect in the dura (the covering the spinal tube)
3. A congenital weakness such as high levels of beta-lipoproteins and cholesterol and muscular dystrophy
4. Spinal surgery and spinal trauma
5. Lumboperitoneal shunt and Ventriculoperitoneal shunt procedures for hydrocephalus
6. Severe dehydration
7. Diabetic coma

8. Uremia is an excess of nitrogenous waste products in the blood and their toxic effects
9. Severe systemic illness such as fever, sweating, weight loss, etc.

A syndrome of ICH is a single pathophysiological entity of diverse origin. Usually, it is characterized by an orthostatic headache, that is, one that occurs or worsens with upright posture, although patients with chronic headaches or even no headache have been described. The nature and location of the headache vary greatly from patient to patient; but consistently the pain is exacerbated by laughing, coughing, jugular venous compression, and Valsalva maneuver, and is resistant to treatment with analgesic agents.

In addition to headache, patients may experience nausea, vomiting, anorexia, neck pain, dizziness, horizontal diplopia, changes in hearing, galactorrhea, facial numbness or weakness, or radicular symptoms involving the upper limb, all of which are orthostatic in nature. Intracranial hypotension generally is considered to be a benign condition, and most cases resolve with conservative management. With advances in diagnosis, however, atypically disabling presentations are increasingly recognized including parkinsonism, frontotemporal dementia, syringomyelia, hypopituitarism, seizures, coma, and death.

Symptoms range from mild to severe headache. Other symptoms can include nausea, vomiting, double vision and difficulty with concentration.

3.18 Transient Global Amnesia

Transient global amnesia (TGA) is a sudden, temporary episode of temporary disturbance in an otherwise healthy person's memory and can't be attributed to a more common neurological condition, such as epilepsy or stroke.

During an episode of transient global amnesia, your recall of recent events simply vanishes, so you can't remember where you are or how you got there. You may also draw a blank when asked to remember things that happened a day, a month or even a year ago. Though patients generally remember their own identities, they are often very confused by their surroundings and the people around them.

There are two kinds of TGA: anterograde and retrograde amnesia. Anterograde amnesia is a type of memory loss associated with a trauma, disease, or emotional events. It is characterized by the inability to remember new information. Retrograde amnesia is associated with the loss of distant memories usually preceding a given trauma.

In transient global amnesia, generally both distant memories and immediate recall are retained, as are language function, attention, visual-spatial and social skills. However, during the period of amnesia, people suffering from the disorder cannot remember recent occurrences nor can they retain any new visual or verbal information for more than a couple minutes. They continuously ask questions about events that are transpiring, for example where they are, who is accompanying them, why they are coming to this place. However, once they are told, they immediate forget the answer, and repeat the question again.

Fortunately, transient global amnesia is rare and has a very positive prognosis - its effects are never permanent and the episodes last for a relatively short period of time, and afterward your memory is fine. However, the inability to remember can be extraordinarily frightening.

TGA generally affects fifty to eighty-year-old men, and patients with certain diseases, for example liver cirrhosis and kidney dysfunction or failure.

Causes of TGA are not known, however, there is convincing evidence that external emotional stresses, such as sexual intercourse, immersion in cold water, or strenuous physical exertion, can trigger the associated loss of memory.

3.19 Neuroaxonal Dystrophies

This is a slowly progressive disorder in which swellings (spheroids) develop and accumulate along axons throughout the brain and spinal cord. Axons are the parts of the nerve cells along which electrical signals travel, and so this condition results in abnormal conduction of nervous impulses, and associated clinical signs such as a high-stepping gait and lack of coordination. Neuroaxonal dystrophy is seen in various genetic diseases, vitamin deficiencies, and aging.

Neuroaxonal dystrophy Inheritance is believed to be autosomal recessive (a mutation in a gene on one of the first 22 nonsex chromosomes can lead to an autosomal disorder, 23rd chromosome is for sexual recessive). Infantile neuroaxonal dystrophy (INAD) is a rare inherited neurological disorder. It affects axons and causes progressive loss of vision, muscular control, and mental skills.

While the basic genetic and metabolic causes are unknown, INAD is the result of an abnormal build-up of toxic substances in nerves that communicate with muscles, skin, and the conjunctive tissue around the eyes. Symptoms usually begin within the first 2 years of life, with the loss of head control and the ability to sit, crawl, or walk, accompanied by deterioration in vision and speech.

Individuals with infantile neuroaxonal dystrophy typically do not have any symptoms at birth, but between the ages of about 6 and 18 months they begin to experience delays in acquiring new motor and intellectual skills, such as crawling or beginning to speak. Eventually they lose previously acquired skills. In some cases, signs and symptoms of infantile neuroaxonal dystrophy first appear later in childhood or during the teenage years and progress more slowly.

Rapid, involuntary eye movements (nystagmus), eyes that do not look in the same direction (strabismus), and vision loss due to deterioration (atrophy) of the optic nerve are characteristic of infantile neuroaxonal dystrophy. Hearing loss may also develop. Children with this disorder experience progressive deterioration of cognitive functions (dementia), and eventually lose awareness of their surroundings.

3.20 Diffuse Cerebral Sclerosis of Schilder

A rare central nervous system demyelinating condition affecting children and young adults. All are characterized by demyelination of the white matter of the brain, with muscle spasticity, optic neuritis, aphasia, cortical deafness, spastic hemiplegia, adrenal insufficiency, pseudobulbar palsy and dementia. Many of the signs resemble those of multiple sclerosis. There is no known treatment. The cause may be viral or genetic. Myelin acts as a shied to the nerve.

Schilder's disease is synonym with poliodystrophia cerebri, encephalitis periaxialis, cerebral sclerosis, diffuse, Alpers' disease, encephalitis periaxialis concentrica, Alpers syndrome, encephalitis periaxialis diffusa, myelinoclastic diffuse sclerosis, and Balo's concentric sclerosis

3.21 Subdural Effusion

A subdural effusion is a rare complication of bacterial meningitis. Subdural effusion is more common in infants and in persons who have meningitis caused by haemophilus influenzae. Symptoms of subdural effusion includes:

- Bulging fontanelles
- Lethargy
- Persistent fever
- Increased head circumference
- Seizures
- Separated sutures
- Vomiting
- Weakness

3.22 Alzheimer Disease

This incurable, degenerative, and terminal disease was first described by German psychiatrist and neuropathologist Alois Alzheimer in 1906 and was named after him.

Alzheimer's disease (AD) is the most common form of dementia among older people. Dementia is a brain disorder that seriously affects a person's ability to carry out daily activities.

AD begins slowly. It first involves the parts of the brain that control thought, memory and language. People with AD may have trouble remembering things that happened recently or names of people they know. Over time, symptoms get worse. People may not recognize family members or have trouble speaking, reading or writing. They may forget how to brush their teeth or comb their hair. Later on, they may become anxious or aggressive, or wander away from home. Gradually, bodily functions are lost, ultimately leading to death. The mean life expectancy following diagnosis is approximately seven years. Eventually, they need total care. This can cause great stress for family members who must care for them.

AD usually begins after age 60. In 2006, there were 26.6 million sufferers worldwide. Alzheimer's is predicted to affect 1 in 85 people globally by 2050. The risk goes up as one gets older. The risk is also higher if a family member has had the disease.

The causes and progression of Alzheimer's disease are not well understood. Research indicates that the disease is associated with plaques and tangles in the brain. Plaques form when protein pieces called beta-amyloid clump together. Beta-amyloid comes from a larger protein found in the fatty membrane surrounding nerve cells. Beta-amyloid is chemically "sticky" and gradually builds up into plaques.

Currently used treatments offer a small symptomatic benefit; no treatments to delay or halt the progression of the disease are as yet available. However, some drugs may help keep symptoms from getting worse for a limited time. As of 2008, more than 500 clinical trials have been conducted for identification of a possible treatment for AD, but it is unknown if any of the tested intervention strategies will show promising results. A number of non-invasive, life-style habits have been suggested for the prevention of Alzheimer's disease, but there is a lack of adequate evidence for a link between these recommendations and reduced degeneration. Mental stimulation, exercise, and balanced diets are suggested, as both a possible prevention and a sensible way of managing the disease, Can Alzheimer's disease be prevented. National Institute on Aging. 2008-02-29. http://www.nia.nih.gov/NR/rdonlyres/63B5A29C-F943-4DB7-91B4-0296772973F3/0/CanADbePrevented.pdf.

Alzheimer's disease is characterized by loss of neurons and synapses in the cerebral cortex and certain subcortical regions. This loss results in gross atrophy of the affected regions, including degeneration in the temporal lobe and parietal lobe, and parts of the frontal cortex and cingulated gyrus. Studies using MRI and PET have documented reductions in the size of specific brain regions in patients as they progressed from mild cognitive impairment to Alzheimer's disease, and in comparison with similar images from healthy older adults, Moan R (July 20, 2009). "MRI software accurately IDs preclinical Alzheimer's disease". *Diagnostic Imaging*.

203

Figure (3.5) shows a comparison of a normal aged brain (left) and an Alzheimer's patient's brain (right).

Figure (3.5): Comparison between a normal brain and patient's brain with Alzheimer's

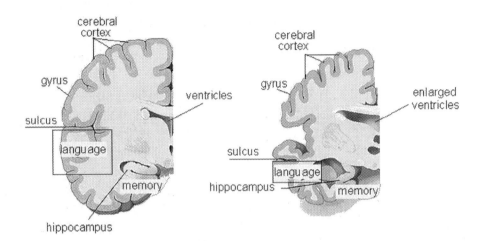

Alzheimer's disease has been linked to genes on three chromosomes - 14, 19, and 21. The apoE4 gene on chromosome 19 has been linked to late-onset Alzheimer's disease, which is the most common form of the disease. One hypothesis suggests that the apoE4 protein leads to the neurofibrillary tangles, Figure (3.6).

Figure (3.6): Accumulation of twisted protein filaments (neurofibrils) within neurons

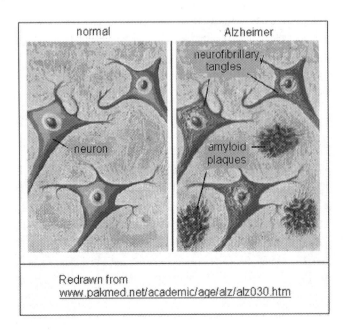

Other risk factors for Alzheimer's disease include high blood pressure (hypertension), coronary artery disease, diabetes, and possibly elevated blood cholesterol. Individuals who have completed less than eight years of education also have an increased risk for Alzheimer's disease. These factors increase the risk of Alzheimer's disease, but by no means do they mean that Alzheimer's disease is inevitable in persons with these factors.

All patients with Down Syndrome will develop the brain changes of Alzheimer's disease by 40 years of age. This fact was also a clue to the "amyloid hypothesis of Alzheimer's disease".

The Alzheimer's Association has developed the following list of warning signs that include common symptoms of Alzheimer's disease.

1. Memory loss
2. Difficulty performing familiar tasks
3. Problems with language
4. Disorientation to time and place
5. Poor or decreased judgment
6. Problems with abstract thinking
7. Misplacing things
8. Changes in mood or behavior
9. Changes in personality
10. Loss of initiative

3.23 Parkinson's Disease

Parkinson's disease (PD) belongs to a group of conditions called motor system disorders, which affects nerve cells, or neurons, in a part of the brain that controls muscle movement. In Parkinson's, neurons that make a chemical called dopamine (dopamine is manufactured in an area of the brain called the substantia nigra) die or do not work properly. Dopamine normally sends signals that help coordinate your movements. No one knows what damages these cells.

Symptoms are:

1. Tremor, or trembling in hands, arms, legs, jaw, and face
2. Rigidity, or stiffness of the arms, legs and trunk;
3. Bradykinesia, or abnormal slowness of movements and reflexes (Bradykinesia was clearly present in the less affected patients with PD, and worsened as the disease severity increased. Hypokinesia [slow or diminished movement of body musculature], however, emerged prominently only in the more affected patients); and
4. Poor balance and coordination.

Parkinson's is not a fatal condition. However, the end stage of the disease can lead to pneumonia, choking, severe depression, sleep problems or trouble chewing, swallowing or speaking and death. As symptoms get worse, people with the disease may have trouble walking, talking or doing simple tasks.

In most cases, researchers don't know the causes of Parkinson diseases. Causes may be due to the followings:

1. Dopamine-producing cells in the substantia nigra are lost or dead. This is due to a protein called *synuclein* which accumulates to form protein deposits called *Lewy bodies*. Researchers believe that Parkinson's disease is a late complication of protein accumulation, where the protein can accumulate in other areas of the brain and in the intestinal tract. This case belongs to the so called *Primary Parkinsonism.*

2. *Secondary Parkinsonism* is due to some disease (e.g., nervous system conditions, heart disease, brain tumours, viruses) or chemical interfering with or damaging dopamine-producing cells in the brainstem. The most common cause is side effects of medication for other problems. Medications that can cause secondary Parkinsonism include:

   haloperidol* and other medications used to treat hallucinations
   metoclopramide (an antinausea medication)

   Less common causes of secondary Parkinsonism include poisoning by carbon monoxide or manganese (a type of mineral), lesions and tumours in the brainstem, and a rare illicit drug called N-MPTP. An outbreak between 1918 and 1924 of a disease called von *Economo's encephalitis* left thousands of people across North America with Parkinson's.

3. A number of genetic mutations have recently been identified, suggesting that Parkinson's may run in some families. However, a major US twin study suggested that environment plays a larger role than inheritance. The current consensus is that genetic factors are dominant only in Parkinson's that appears before age 50.

3.24 Schizophrenia and Catatonic Schizophrenia

Schizophrenia is a serious mental illness characterized by a disintegration of the process of thinking and of emotional responsiveness. It most commonly manifests as auditory hallucinations, paranoid or bizarre delusions, or disorganized speech and thinking with significant social or occupational dysfunction. Hallucinations most often consist of hearing voices that comment on behaviour, are insulting or give commands. Less often, people with schizophrenia may see or feel things that aren't there.

People with schizophrenia may hear voices other people don't hear. They may believe other people are reading their minds, controlling their thoughts, or plotting to harm them. This can terrify people with the illness and make them withdrawn or extremely agitated.

People with schizophrenia may not make sense when they talk. They may sit for hours without moving or talking. Sometimes people with schizophrenia seem perfectly fine until they talk about what they are really thinking.

Catatonic schizophrenia is one of several types of schizophrenia, a chronic mental illness in which reality is interpreted abnormally (psychosis). Catatonic schizophrenia includes extremes of behavior. At one extreme of catatonic schizophrenia, you're unable to speak, move or respond. At the other, you have overexcited or hyperactive motion and you may involuntarily imitate sounds or movements of others.

The symptoms of catatonic schizophrenia include one or more of the following categories:

1. Physical immobility – This includes inability to move purposefully within physical environment, reluctance to attempt movement, limited range of movement, decrease muscle strength, and inability to perform action as instructed. Patients also may have a form of immobility known as waxy flexibility. For example, if your arm is moved into a certain position, it will stay in that position for hours.

2. Excessive mobility – Some people with catatonic schizophrenia show excessive and purposeless mobility that is not influenced by external stimuli. They may whirl their arms, pace rapidly and in a frenzy manner, or make loud noises.

3. Extreme resistance – Patients show extreme resistance and considerable strength in resisting any attempt to change their posture (rigidity) or may not speak at all.

4. Peculiar movements – The patient willingly takes on a bizarre or inappropriate stance or posture, or engages in peculiar movements, pronounced grimacing or unusual mannerisms for long periods. The patient will repeat words, obsessively follow a routine, or always arrangeobjects exactly the same way.

5. Mimicking speech or movement – Patients are overexcited or hyperactive, sometimes mimicking sounds (echolalia) or movements (echopraxia) around them - often referred to as catatonic excitement. Patients may also present other disturbances of movement - seemingly purposeless actions are performed repetitively (stereotypic behavior), sometimes to the exclusion of involvement in any creative or productive activity.

6. Other signs and symptoms of catatonic schizophrenia

Although the main symptoms of catatonic schizophrenia are catatonic behaviors, you may also have some of the other common signs and symptoms of schizophrenia, such as:

- Seeing or hearing things that don't exist (hallucinations), especially voices
- Lack of emotions
- Having beliefs not based on reality (delusions)
- Angry outbursts
- Neglect of personal hygiene
- Emotions inappropriate to the situation
- Incoherent speech
- Trouble functioning at school or work
- Social isolation
- Clumsy, uncoordinated movements

3.24.1 Treatment of Schizophrania

Medications called neuroleptics or antipsychotic drugs have proven to be among the most important medical advances for treating schizophrenia. Because of these drugs, people with schizophrenia no longer need to be hospitalized for years. Most are able to live in the community, needing hospitalization for the illness only if they relapse.

Antipsychotic drugs are the best treatment now available, but they do not "cure" schizophrenia or ensure that there will be no further psychotic episodes. The choice and dosage of medication can be made only by a qualified physician who is well trained in the medical treatment of mental disorders. The dosage of medication is individualized for each patient, since people may vary a great deal in the amount of drug needed to reduce symptoms without producing troublesome side effects.

Psychotherapy is also widely recommended and used in the treatment of schizophrenia, although services may often be confined to pharmacotherapy because of reimbursement problems or lack of training.

Cognitive Behavioral Therapy (CBT) is used to target specific symptoms and improve related issues such as self-esteemed, social functioning, and insight. Although the results of early trials were inconclusive as the therapy advanced from its initial applications in the mid 1990s, more recent reviews clearly show CBT is an effective treatment for the psychotic symptoms of schizophrenia, Cormac I, Jones C, Campbell C (2002). "Cognitive behaviour therapy for schizophrenia". *Cochrane Database of systematic reviews* (1): CD000524.

Tobacco smoking has been proven to ease effects of schizophrenia; it has been proposed that thenicotine patch is a treatment for schizophrenia.

3.25 Creutzfeldt-Jacob Disease

Creutzfeldt-Jakob Disease (CJD) is a rare, degenerative neurological disorder (brain disease). It is incurable and invariably a fatal brain disorder. It affects about one person in every one million people per year worldwide. In the United States there are about 200 cases per year. CJD usually appears in later life and runs a rapid course. Typically, onset of symptoms occurs about age 60, and about 90 percent of individuals die within 1 year. It is a type of transmissible spongiform encephalopathy (TSE) which is caused by prion. Prion proteins occur in both a normal form, which is a harmless protein found in the body's cells, and in an infectious form, which causes disease. The word prion, coined in 1982 by Dr. Stanley B. Prusiner, is a portmanteau derived from the words *protein* and *infection*. Prions are the cause of a number of diseases in a variety of mammals, including bovine spongiform encephalopathy (BSE, also known as "mad cow disease") in cattle and CJD in humans.

Spongiform encephalopathy refers to the characteristic appearance of infected brains, which become filled with holes until they resemble sponges under a microscope. CJD is the most common of the known human TSEs . Other human TSEs include kuru, fatal familial insomnia (FFI), and Gerstmann-Straussler-Scheinker disease (GSS). Kuru was identified in people of an isolated tribe in Papua New Guinea and has now almost disappeared. FFI and GSS are extremely rare hereditary diseases, found in just a few families around the world.

As the illness progresses, mental deterioration becomes pronounced and involuntary movements, blindness, weakness of extremities, and coma may occur. Initially, individuals experience problems with muscular coordinationand personality changes, including impaired memory, judgment, and thinking; and impaired vision. Pneumonia and other infections often occur in these individuals and can lead to death.

There are two types of CJD; classical and variant.

There are three major categories of classical CJD:

- In sporadic CJD, the disease appears even though the person has no known risk factors for the disease. It occurs for no known reason.This is by far the most common type of CJD and accounts for at least 85 percent of cases.

- Hereditary CJD runs in the family. The patient has a family history of the disease and/or tests positive for a genetic mutation associated with CJD. About 5 to 10 percent of cases of CJD in the United States are hereditary.

- In acquired CJD, the disease is transmitted by contact with infected tissue or exposure to brain or nervous system tissue, usually during certain medical procedures. There is no evidence that CJD is contagious through casual contact with a CJD patient. Since CJD was first described in 1920, fewer than 1 percent of cases have been acquired CJD.

Variant CJD symptoms include early psychiatric symptoms such as anxiety, depression, withdrawal and behavioural changes. Persistent pain or odd sensations in the face or limbs often develop. The disease then progresses to motor difficulties, involuntary movements and mental deterioration, often ending in a persistent vegetative state. The patient may live on average for about one year after the onset of symptoms.

At present, there is no cure or treatment to slow the progression of the disease.

3.26 Encephalitis Lethargica

Encephalitis lethargica is a disease characterized by high fever, headache, double vision, delayed physical and mental response, and lethargy. In acute cases, patients may enter coma. Patients may also experience abnormal eye movements, upper body weakness, muscular pains, tremors, neck rigidity, and behavioral changes including psychosis. The cause of encephalitis lethargica is unknown. Between 1917 to 1928, an epidemic of encephalitis lethargica spread throughout the world, but no recurrence of the epidemic has since been reported. Post encephalitic Parkinson's disease may develop after a bout of encephalitis-sometimes as long as a year after the illness.

Encephalitis lethargica is an inflammation of the brain caused by two trypanosomes (microscopic protozoan parasites). The illness, which can be fatal, is transmitted from one infected person to another by the tsetse fly. While it can occur globally, encephalitis lethargica is especially prevalent in Africa. As of 2004, the disease is a threat to more than 60 million people in 36 sub-Saharan African countries. In 1999, nearly 45,000 cases were reported, according to the World Health Organization (WHO).

Appendix - 1

Case Study Using Stanford-Binet Scale Test

The co-author of this book (Dima Elsersawi) used Stanford-Binet Intelligence scale to evaluate the maternal affect attunement on their toddlers. In her method, there were thirty-one mother child dyads participated in the study which was conducted at the Infancy Research Center at York University, Toronto, Ontario, Canada. Twenty-four of the mother-child dyads were part of a longitudinal study designed to examine the relationship of maternal affect attunement and children's internal state language. Seven cross-sectional mother-child dyads were also included in the study and were recruited by telephone from birth announcement archives of the Toronto Star Newspaper. Since the sample was part of a larger project that investigated other variables that are irrelevant to this particular study, seven babies were disqualified: 2 for failing to complete the intelligence test, 3 for fussiness, and 2 for experimental error. The children were healthy 31-34 months old. Eighty-four percent of participating children were either first-or-second born.

The children came from middle-class families, indicated by an average of three years and nine months of parental post-secondary education. Ethnic breakdown of the sample revealed that 62.5% of the participating were Caucasian, 12.5% were Asian, 0.4% were African-Canadian and 20.8% were of mixed background. Fifty-eight percent of the participants indicated that the English was the language spoken at home, while forty-two percent spoke another language.

Participant received information about the study in the mail following telephone solicitation, and they received a second call several days later to see if they were interested in participating in the study. Upon agreement, participants were seen on two different occasions. The first visit was at the laboratory and the second was at home.

Procedures

Visit 1. Upon arrival in the lab, the purpose and procedure of the study were explained. Mothers were then asked to read and sign consent forms and as well to fill a participant information sheet. Mothers and children were asked to sit on a rug where they were observed for a duration of 10 minutes of free play. Subjects were instructed to ignore the camera and to play with their infants just as they normally would at home. The are was equipped with age-appropriate toys (a doll, a jack in the box, a panda hand-puppet, a drum and a cash register). At the end of the 10 minutes, the mother-child dyads were offered snacks (cookies, crackers, and juice) while still playing for another 5 minutes. All along, an experimenter was videotaping a total of 15 minutes of mother-child interaction.

<u>TV Situation</u>. Mother and child were requested to sit on a rug to watch four minutes of a series of video clips of babies displaying happy, sad, and neutral expressions on television. The clips are of infants filmed during a previous study at the Infancy Centre at York University. The mother was instructed to interact with her child as she normally would while watching TV at home. She was asked not to hold her child's arm or hands and to refrain from pointing to the TV. If the child got up and wondered around, the mother was requested to ask the child to sit back on the rug.

The clips were randomized based on a Latin-square design (Sharma, 1975) and copied on four separate videotapes. The order of these presentations was randomized as well as the TV and the structured play situations.

<u>Structured Play Situation</u>. Mothers and toddlers were seated at a table where they played with small-sized doll in happy, sad and neutral stuations. Because performance on one situation may have an effect on the other, the order of presentation of the three tasks was always counterbalanced. The mother was requested to play with her child as she normally would for 3 minutes each situation, totaling 9 minutes of play interaction. Prior to each situation, an appropriate script was verbally presented to each dyads. After the end of each situation, the research assistant would enter the room and give thetoys and the script of the next situation.

A description of the three structured situations is as follows:

A) Neutral situation: The dyad was instructed to create a neutral situation by putting the baby to sleep. They were to prepare for bed, change the diapers, put on pajamas and brush the baby's hair. The mothers were asked to be as neutral as possible.

B) Sad situation: Mothers were requested to emphasize the sadness of the situation as much as they could. The experimenter would say, "it is a very sad day. The baby fell down and broke a leg and an arm. Baby is at hospital right now and in real pain. The baby feels very sick". The mother and child were asked to put a bandage on the baby, to consult the x-ray and to check the baby's temperature. In order to prevent mothers from turning into a happy story after the treatment of the baby, mothers were requested to maintain a sad theme throughout the situation.

C) Happy situation: This is a happy theme since it is the birthday of the baby. Mother and toddler were requested to prepare the baby for the party, to dress the baby, to prepare the cake, to play games and to open the present. Upon completion of the task, mothers and toddlers were brought into a strange room in the lab for a final task to assess attachment.

The Strange Situation: Infant attachment to the mother was assessed using this paradigm, which took approximately 25 minutes to complete. The strange situation (Ainsworth, Blehar, Waters, & Wall, 1978), consists of eight episodes of increasing stress for the infant, including the entrance of unfamiliar adult and two departures by the mother. After the first separation, the mother was instructed to rejoin the stranger (who left immediately after the mother's entrance) and child in the room, pause for about 10 seconds to give to give the child a chance to react to her entrance, and then do whatever she wants. The mother was told she could hold and comfort her child if necessary and to engage the child in playing with the toys again. During this paradigm, the child's emotional arousal was heightened to necessitate mother's sensitivity, and both mother and child were more likely to exhibit a range of positive and negative behaviours. However, the Strange Situation is incorporated for the purpose of the larger study and is not relevant to this particular study, thus involved behaviours will not be further analyzed.

Prior to leaving the lab, the mothers were asked to schedule a home visit for times when they felt that their infants were rested and alert. To ensure that toddlers were at approximately the same level of receptive and productive vocabulary, their mothers were asked to take home the words and Gesture Scale of the MacArthur Communicative Development Inventories (CDI) (Fenson, Dale, Reznick, Bates, Thal&Pethick. 1994).

Visit 2. The objective of the second visit was to examine the mother's interaction with her child through observation of the two in a naturalistic setting around the home, and to administer intelligence tests to the dyads. Upon arrival, researchers explained the nature and the procedure of the home study. Since a cooperative relationship between the examiner of the intelligence test and the child is essential for a valid assessment, 7 to 10 minutes of free play between the two was spent at the beginning of the visit. Another 7 to 10 minutes of interplay was given to the mother and child. The dyads were instructed to play like they normally would while being observed and videotaped.

Upon completion of the interaction, the experimenters simultaneously administered the intelligence tests to the mother and child, each in separate areas. The Wechsler Abbreviated Scale of Intelligence (WASI) was given to the mother and Stanford 4th edition was administered to the child. Half way through the tests the mother and child were given a snack break, which was also videotaped to assess their interaction.

Measures

Relevant measures essential for examining the relationship between maternal affect attunement, maternal intelligence, and toddler IQ, will be discussed. The other variables are used for the investigation of the larger project and are not relevant to the purpose of this study.

<u>Wechsler Abbreviated Scale of Intelligence (WASI)</u>: this test is administered to people ages 6 through 89 years. The four subsets of the WASI measure both fluid and crystallized intelligence. The Block Design and Matrix Reasoning subset assess performance and measures fluid ability, whereas the Vocabulary and Similarities subsets assess vocabulary scores and measures crystallized intelligence. To alternate between Verbal and Performance subsets, the tests are administered in the following sequence: Vocabulary, Block Design, Similarities, Matrix Reasoning. The sequence is preferred because it alternates Verbal and Performance subsets, and the picture items in the Vocabulary section will help the examinee "warm up" for the rest of the test. Scoring rules and criteria are provided for each subset.

In most cases, scoring is objective; however, for Vocabulary and Similarities, more judgment is required. For these subsets, participants' responses are compared to sample responses that are provided in addition to the general scoring criteria. 0, 1 or 2 points are rewarded for each response.

<u>A description of the subsets is as follows:</u>

Vocabulary: This subset includes 4 picture items and 38 word items. The mother is asked to give an oral definition of the words. There are no stringent time limits but the examinee is not allowed more than approximately 30 second per item.

Block Design: For this subset, the mother uses blocks to replicate two colour designs within the specified time limit. The 13 designs progress in difficulty from simple designs requiring two blocks to more complex designs nine blocks.

Similarities: The mother is read two words that present common concepts or objects. Her task is to state how the two words are alike.

Matrix Reasoning: This subset is composed of four types of nonverbal reasoning tasks: Pattern completion, classification, analogy, and serial reasoning. The mother is asked to look at a matrix from which a section is missing and complete the matrix either by saying the number of or by pointing to one of five response options. There are no time limits fro this subset, however, if the examinee does not finish an item within 30 seconds, the examinee should proceed to the next item.

Stanford-Binet Intelligence Scale (SBIS), 4th edition: This scale is used to assess intelligence for children between 2 and 18 years of age and provides intelligence estimates up to 23 years of age. In this study, testing was done on 4 subsets short form because of its relatively short administration time. The 4 subsets included:

Vocabulary: Picture naming and vocabulary definition.

Pattern Analysis: Block placement and block design tasks.

Quantitative: This test proceeds from placing blocks with varying numbers of dots correctly on a tray to counting the numbers of children, pencils, etc. on cards, including simple subtraction, to relations, to arithmetic illustrated with pictures, to verbally enclosed arithmetic questions.

Bead memory: This test proceeds from pointing out the correct colored and shaped bead in a box as (shown on a card) to arranging beads on a stick with base after the pictured pattern has been exposed for five seconds.

Of the four administered tests, only pattern analysis had definite time limits, while for other tests it was left to the examiner to determine whether a satisfactory response can be elicited by allowing more time. Vocabulary is given first as a "routing test" that determines the items to begin with on each test. A "basal level" is then established for each test (passing two consecutive items and testing is continued until the ceiling level (four consecutive failures on each test) is reached). The test items are presented in a standardized manner, but some flexibility is acceptable; for example the examiner can shift to another test if the child is fatigue or resistant to a given test.

At the end of the visit, mothers were asked to return the MacArthur Communicative Development Inventories (CDI), which was given to them on the first visit at the lab.

Measuring Maternal Interactive Style

Maternal interactive style was assessed using 10 minutes of the total mother-child free play interaction as well as 10 minutes of the total snack time during both visit 1 and visit 2. The first minute of each free play interaction was not to be coded. An internal time generator on a VCR was used to impose a timeline on the remaining minutes of video-recording.

Interactive style was measured using a coding system adapted in part from Legerstee and Varghese (2001) and Ainsworth et al., (1971). Three variables were coded to reflect the quality of maternal interactive styles: (1) maintaining attention, (2) warm sensitivity and (3) responsiveness. A scale of 1 to 9 (lowest to highest) was used fro each of four sessions to score the mothers on the above variables. An average score was taken for each mother and a median split was used to categorize mothers into two clusters (HA and LA).

Maintaining attention is defined as a maternal request, question, or comment related to any activity or object with which the child is currently visually or physically engaged with, or both. Maintaining could also be a maternal request, question, or comment that is in direct response to the child's attempt to attract the mother's attention to an object or activity.

Warm sensitivity is a composite assessment of the degree of sensitivity that mothers display to their children's affective cues, including promptness and appropriateness of reactions, acceptance of the children's interests, amount of physical affection, positive affect, and tone of voice.

Maternal responsive is defined as imitative response to children's smiles and vocalizations, and as modulations of children's negative affect. The scoring procedure for this bahaviour involved going back to each of the infants' initiation of social contact (smiles, vocalizations, gestures) and observing for maternal response to the infants' smile or laugh, or accepting a toy offered by the infant.

Inter-rater Reliability

Three coders coded the mothers' interactive bahaviours. In order to establish reliability, all three-research assistants randomly coded the same 20% of the data. Each partner independently coded one third of the remaining 80% of the data. Cohen's Kappa coefficient was computed separately for each condition. (Kappa scores were used because they correct for chance). Agreement was assessed for all behaviours. Training continued until Kappa scores reached 0.90 fro each partner with regard to maternal interactive style.

RESULTS

Maternal Affect Attunement

In order to divide mothers into two groups: High Affect Attunement (HA) or Low Affect Attunement (LA), their scores were rank-ordered from lowest to highest. A median split was used to divide them into either HA or LA categories. A median split was used to divide them into either HA or LA categories. A scale of 1 to 9 (1 = lowest, 9 = highest) was used to rate mothers on affect attunement. The score ranged from 2.25 (LA) to 8.75 (HA) (M = 6.24, SD = 1.74).

Maternal Intelligence

In order to classify mothers into two groups (High or Low IQ), their full IQ scores on the Wechsler Abbreviated Scale of Intelligence (WASI) were rank-ordered from lowest to highest. A median split was used to divide the categories. The scores ranged from 83 to 137 (M = 108.62, SD = 10.99).

Toddler Intelligence

The children's total composite score on Stanford-Binet Scale of Intelligence (SBIS) ranged from 99 to129 (M = 114.91), SD = 8.31).

<u>Relation between Maternal Affect Attunement and Toddler Intelligence</u>

A correlational analysis was conducted to determine whether there was a relation between an affectively attuned maternal interactive style and toddler intelligence. Results of this analysis showed an insignificant relationship between those variables (r = -.119, p = .580). The means of child IQ for the two groups (HA) and (LA) were the same, including no main effect of maternal interactive style on toddler IQ. (HA, M = 114.91, SD = 8.88), (LA, M = 114.91, SD = 8.07).

<u>Relation between Maternal Intelligence and Toddler Intelligence</u>

To find out the relation between maternal IQ and toddler IQ, a correlational analysis was performed. Again, the results were insignificant (r = .258, p = .223). However, the means of child IQ for the two groups (High IQ and Low IQ mothers) were different, indicating a main effect of maternal intelligence on child intelligence. (High IQ, M = 116.85, SD = 7.72).

<u>The Effect of Maternal High IQ and HA on Toddler IQ</u>

To determine whether children whose mothers scored high IQ and HA will exceed all other children on intelligence, a correlational analysis was conducted. Although, the finding was insignificant (r = .084, p = .698), the mean (M = 120.00, SD = 8.06) of this cohort of toddlers was higher than all other groups, signifying that the combination of high maternal intelligence and HA have a stronger impact on toddler intelligence than each variable alone.

Conclusion

The present results suggest that maternal intelligence plays a stronger role than affect attunement in the child's intellectual capacities, however interpretation of the results is limited by the fact that the analysis do not address the continuity of maternal interactive behavior nor do they examine the stability of intelligence in a longitudinal approach.

The implications of the study suggest some direction for future research. Careful investigation of consistencies and inconsistencies in maternal affect attunement and their relationship with child IQ at later points in their life span seem to be important for future examination.

Future studies should also employ a longitudinal approach in measuring child intelligence in order to study the relationship between maternal IQ and child IQ.

Glossary

Acquired Epileptiform Aphasia (AEA) or Landau-Kleffner Syndrome (LKS): Landau-Kleffner syndrome (LKS) is a rare, childhood neurological disorder characterized by the sudden or gradual development of aphasia (the inability to understand or express language) and an abnormal electro-encephalogram (EEG). LKS affects the parts of the brain that control comprehension and speech.

Acute Disseminated Encephalomyelitis: Acute disseminated encephalomyelitis (ADEM) is characterized by a brief but intense attack of inflammation in the brain and spinal cord that damages myelin – the protective covering of nerve fibers.

Adie Syndrome: Adie-Holmes syndrome (HAS) is a neurological disorder affecting the pupil of the eye and the autonomic nervous system. It is characterized by one eye with a pupil that is larger than normal and constricts slowly in bright light (tonic pupil), along with the absence of deep tendon reflexes, usually in the Achilles tendon.

Adrenoleukodystrophy: Adrenoleukodystrophy (ALD) is one of a group of genetic disorders called the *leukodystrophies* that cause damage to the myelin sheath, an insulating membrane that surrounds nerve cells in the brain.

Agenesis of the Corpus Callosum: Agenesis of the corpus callosum (ACC) is a birth defect in which the structure that connects the two hemispheres of the brain (the corpus callosum) is partially or completely absent.

Agnosia: Agnosia is a rare disorder characterized by an inability to recognize and identify objects or persons.

Alexander Disease: Alexander disease is one of a group of neurological conditions known as the leukodystrophies, disorders that are the result of abnormalities in myelin, the "white matter" that protects nerve fibers in the brain.

Alpers' Disease: Alpers' disease is a rare, genetically determined disease of the brain that causes progressive degeneration of grey matter in the cerebrum.

Alzheimer's Disease: Alzheimer's disease (AD) is an age-related, non-reversible brain disorder that develops over a period of years. Initially, people experience memory loss and confusion, which may be mistaken for the kinds of memory changes that are sometimes associated with normal aging.

Amyotrophic Lateral Sclerosis: Amyotrophic lateral sclerosis (ALS), sometimes called Lou Gehrig's disease, is a rapidly progressive, invariably fatal neurological disease that attacks the nerve cells *(neurons)* responsible for controlling voluntary muscles. In ALS, both the upper motor neurons and the lower motor neurons degenerate or die, ceasing to send messages to muscles.

Anencephaly: Anencephaly is a defect in the closure of the neural tube during fetal development. The neural tube is a narrow channel that folds and closes between the 3rd and 4th weeks of pregnancy to form the brain and spinal cord of the embryo.

A cerebral Aneurysm: A cerebral Aneurysm is the dilation, bulging, or ballooning-out of part of the wall of an artery in the brain.

Anoxia or Cerebral Hypoxia: Anoxia refers to a condition in which there is a decrease of oxygen supply to the brain even though there is adequate blood flow.

Aphasia: Aphasia is a neurological disorder caused by damage to the portions of the brain that are responsible for language.

Apraxia: Apraxia (called "dyspraxia" if mild) is a neurological disorder characterized by loss of the ability to execute or carry out skilled movements and gestures, despite having the desire and the physical ability to perform them.

Arachnoid Cysts: Arachnoid cysts are cerebrospinal fluid-filled sacs that are located between the brain or spinal cord and the arachnoid membrane, one of the three membranes that cover the brain and spinal cord.

Arachnoiditis: Arachnoiditis describes a pain disorder caused by the inflammation of the arachnoid, one of the membranes that surround and protect the nerves of the spinal cord.

Arnold-Chiari Malformation: Aenold Chiari malformations: (ACMs) are structural defects in the cerebellum, the part of the brain that controls balance.

Asperger Syndrome: Asperger Syndrome (AS) is a developmental disorder. It is an autism spectrum disorder (ASD), one of a distinct group of neurological conditions characterized by a greater or lesser degree of impairment in language and communication skills, as well as repetitive or restrictive patterns of thought and behavior.

Ataxia: Ataxia often occurs when parts of the nervous system that control movement are damaged. People with ataxia experience a failure of muscle control in their arms and legs, resulting in a lack of balance and coordination or a disturbance of gait. The incoordination consists of irregularities in the rhythm, rate, and amplitude of voluntary movements => voluntary movements become jerky and erratic.

Attention Deficit-Hyperactivity Disorder: Attention deficit-hyperactivity disorder (ADHD) is a neurobehavioral disorder that interferes with a person's ability to stay on a task and to exercise age-appropriate inhibition (cognitive alone or both cognitive and behavioral).

Autism: Autism is characterized by three distinctive behaviors. Autistic children have difficulties with social interaction, display problems with verbal and nonverbal communication, and exhibit repetitive behaviors or narrow, obsessive interests.

Autonomic Dysfunction or Dysautonomia : Autonomic dysfunction (AD) is a disorder of autonomic nervous system (ANS) function which is a reflex sympathetic dystrophy, or generalized, as in pure autonomic failure.

Bell's Palsy: Bell's Palsy is a form of temporary facial paralysis resulting from damage or trauma to one of the facial nerves.

Benign Intracranial Hypertension: It means "false brain tumor". It is likely due to high pressure within the skull caused by the buildup or poor absorption of cerebrospinal fluid (CSF).

Binswanger's Disease: Binswanger's disease (BD), also called *subcortical vascular dementia*, is a type of dementia caused by widespread, microscopic areas of damage to the deep layers of white matter in the brain.

Brain and Spinal Tumors: Brain and spinal cord tumors are abnormal growths of tissue found inside the skull or the bony spinal column, which are the primary components of the central nervous system (CNS).

Brain Aneurysm: A cerebral aneurysm is the dilation, bulging, or ballooning-out of part of the wall of an artery in the brain.

Brown-Sequard Syndrome: Brown-Sequard syndrome (BSS) is a rare neurological condition characterized by a lesion in the spinal cord which results in weakness or paralysis (hemiparaplegia) on one side of the body and a loss of sensation (hemianesthesia) on the opposite side.

Bulbospinal Muscular Atrophy: Bulbospinal muscular atrophy is an inherited motor neuron disease that affects males. It is one of a group of disorders called *spinal muscular atrophy* (SMA). Onset of the disease is usually between the ages of 20 and 40, although it has been diagnosed in men from their teens to their 70s. Early symptoms include tremor of the outstretched hands, muscle cramps with exertion, and fasciculations (fleeting muscle twitches visible under the skin). Eventually, individuals develop limb weakness which usually begins in the pelvic or shoulder regions.

CADASIL (Cerebral Autosomal Dominant Arteriopathy with Sub-cortical Infarcts and Leukoencephalopathy): It is an inherited form of cerebrovascular disease that occurs when the thickening of blood vessel walls blocks the flow of blood to the brain.

Canavan Disease: Canavan disease one of the most common cerebral degenerative diseases of infancy, is a gene-linked, neurological birth disorder in which the white matter of the brain degenerates into spongy tissue riddled with microscopic fluid-filled spaces.

Cavernous Malformation: A cerebral cavernous malformation (CCM) is a collection of small blood vessels (capillaries) in the central nervous system (CNS) that is enlarged and irregular in structure.

Central Cervical Cord Syndrome: Central cord syndrome is a form of incomplete spinal cord injury characterized by impairment in the arms and hands and to a lesser extent in the legs.

Cerebellar Degeneration: Cerebellar degeneration is a process in which neurons in the cerebellum - the area of the brain that controls coordination and balance - deteriorate and die.

Cerebellar Hypoplasia: Cerebellar hypoplasia is a neurological condition in which the cerebellum is smaller than usual or not completely developed.

Cerebral Atrophy: Cerebral Atrophy means loss of cells. In brain tissue, atrophy describes a loss of neurons and the connections between them.

Cerebral Beriberi: Cerebral Beriberi is a degenerative brain disorder caused by the lack of thiamine (vitamin B1). It may result from alcohol abuse, dietary deficiencies, prolonged vomiting, eating disorders, or the effects of chemotherapy.

Cerebral Gigantism: Cerebral Gigantism is a rare genetic disorder characterized by excessive physical growth during the first few years of life. Children with Sotos syndrome tend to be large at birth and are often taller, heavier, and have larger heads (macrocrania) than is normal for their age.

Cerebral Hypoxia: Cerebral hypoxia refers to a condition in which there is a decrease of oxygen supply to the brain even though there is adequate blood flow.

Cerebral Palsy: Cerebral Palsy refers to any one of a number of neurological disorders that appear in infancy or early childhood and permanently affect body movement and muscle coordination but don't worsen over time.

Cerebro-Oculo-Facio-Skeletal Syndrome: Cerebro-oculo-facio-skeletal syndrome (COFS) is a pediatric, genetic, degenerative disorder that involves the brain and the spinal cord. It is characterized by craniofacial and skeletal abnormalities, severely reduced muscle tone, and impairment of reflexes.

Chorea: Chorea is an abnormal involuntary movement disorder, one of a group of neurological disorders called *dyskinesias*, which are caused by overactivity of the neurotransmitter dopamine in the areas of the brain that control movement.

Cockayne Syndrome Type II: Cerebro-oculo-facio-skeletal syndrome (COFS) is a pediatric, genetic, degenerative disorder that involves the brain and the spinal cord. It is characterized by craniofacial and skeletal abnormalities, severely reduced muscle tone, and impairment of reflexes.

Colpocephaly: Colpocephaly is a congenital brain abnormality in which the occipital horns - the posterior or rear portion of the lateral ventricles (cavities) of the brain -- are larger than normal because white matter in the posterior cerebrum has failed to develop or thicken.

Coma: A coma, sometimes also called persistent vegetative state, is a profound or deep state of unconsciousness. Persistent vegetative state is not brain-death. An individual in a state of coma is alive but unable to move or respond to his or her environment.

Congenital Myasthenia: All forms of myasthenia are due to problems in the communication between nerve cells and muscles, which involve the activities of neurotransmitters.

Corticobasal Degeneration: Corticobasal degeneration is a progressive neurological disorder characterized by nerve cell loss and *atrophy* (shrinkage) of multiple areas of the brain including the cerebral cortex and the basal ganglia.

Craniosynostosis: Craniosynostosis is a birth defect of the brain characterized by the premature closure of one or more of the fibrous joints between the bones of the skull (called the cranial sutures) before brain growth is complete.

Creutzfeldt-Jakob Disease: Creutzfeldt-Jakob disease (CJD) is a rare, degenerative, invariably fatal brain disorder. Typically, onset of symptoms occurs at about age 60. There are three major categories of CJD: sporadic CJD, hereditary CJD, and acquired CJD.

Dandy-Walker Syndrome: Dandy-Walker syndrome is a congenital brain malformation involving the cerebellum (an area at the back of the brain that controls movement) and the fluid-filled spaces around it.

Dawson Disease: Dawson disease is a progressive neurological disorder of children and young adults that affects the central nervous system (CNS).

De Morsier's Syndrome: De Morsier's syndrome is a rare disorder characterized by abnormal development of the optic disk, pituitary deficiencies, and often agenesis (absence) of the septum pellucidum (the part of the brain that separates the anterior horns or the lateral ventricles of the brain).

Deep Brain Stimulation: Deep brain stimulation (DBS) is a surgical procedure used to treat a variety of disabling neurological symptoms—most commonly the debilitating symptoms of Parkinson's disease (PD), such as tremor, rigidity, stiffness, slowed movement, and walking problems.

Dementia: Dementia is not a specific disease. It is a descriptive term for a collection of symptoms that can be caused by a number of disorders that affect the brain. People with dementia have significantly impaired intellectual functioning that interferes with normal activities and relationships. They also lose their ability to solve problems and maintain emotional control, and they may experience personality changes and behavioral problems, such as agitation, delusions, and hallucinations.

Dyssynergia Cerebellaris Myoclonica: Dyssynergia cerebellaris myoclonica refers to a collection of rare, degenerative, neurological disorders characterized by epilepsy, cognitive impairment, myoclonus, and progressive ataxia.

Developmental Dyspraxia: Developmental dyspraxia is a disorder characterized by impairment in the ability to plan and carry out sensory and motor tasks.

Devic's Syndrome: Devic's syndrome or Neuromyelitis optica (NMO) is an uncommon disease syndrome of the central nervous system (CNS) that affects the optic nerves and spinal cord.

Dravet Syndrome: Dravet syndrome, also called severe myoclonic epilepsy of infancy (SMEI), is a severe form of epilepsy.

Dysarthria: Dysarthria-cerebellar is usually of the spluttering-staccato type, and characterized by slow and slurred with unusual rhythms, and has inappropriate emphases of pitch and loudness, and articulatory impreciseness.

Dysautonomia: Dysautonomia refers to a disorder of autonomic nervous system (ANS) function. Most physicians view dysautonomia in terms of failure of the sympathetic or parasympathetic components of the ANS, but dysautonomia involving excessive ANS activities also can occur.

Dysgraphia: Dysgraphia is a neurological disorder characterized by writing disabilities.

Dyslexia: Dyslexia is a brain-based type of learning disability that specifically impairs a person's ability to read.

Dysmetria: Dysmetria is an inability to place and position a limb correctly, in both range and direction, across the plane of more than one joint.

Dysphagia: Dysphagia is a symptom that accompanies a number of neurological disorders.

Dyspraxia: Dyspraxia is a disorder characterized by an impairment in the ability to plan and carry out sensory and motor tasks.

Dystonias: Dystonias are movement disorders in which sustained muscle contractions cause twisting and repetitive movements or abnormal postures.

Empty Sella Syndrome: Empty Sella Syndrome (ESS) is a disorder that involves the *sella turcica*, a bony structure at the base of the brain that surrounds and protects the pituitary gland.

Encephalitis: Meningitis and encephalitis are inflammatory diseases of the membranes that surround the brain and spinal cord and are caused by bacterial or viral infections.

Encephaloceles: Encephaloceles are rare neural tube defects characterized by sac-like protrusions of the brain and the membranes that cover it through openings in the skull.

Encephalopathy: Encephalopathy is a term for any diffuse disease of the brain that alters brain function or structure.

Epilepsy: Epilepsy is a brain disorder in which clusters of nerve cells, or neurons, in the brain sometimes signal abnormally. In epilepsy, the normal pattern of neuronal activity becomes disturbed, causing strange sensations, emotions, and behavior or sometimes convulsions, muscle spasms, and loss of consciousness.

Fahr's Syndrome: Fahr's syndrome is a rare, genetically dominant, inherited neurological disorder characterized by abnormal deposits of calcium in areas of the brain that control movement, including the basal ganglia and the cerebral cortex.

Familial Hemangioma: A cerebral cavernous malformation (CCM) is a collection of small blood vessels (capillaries) in the central nervous system (CNS) that is enlarged and irregular in structure.

Febrile Seizures: Febrile seizures are convulsions brought on by a fever in infants or small children.

Fibromuscular Dysplasia: Fibromuscular dysplasia (FMD) is the abnormal development or growth of cells in the walls of arteries that can cause the vessels to narrow or bulge. The carotid arteries, which pass through the neck and supply blood to the brain, are commonly affected.

Friedreich's Ataxia: Friedreich's ataxia is an inherited disease that causes progressive damage to the nervous system resulting in symptoms ranging from muscle weakness and speech problems to heart disease.

Frontotemporal Dementia: Frontotemporal dementia (FTD) describes a clinical syndrome associated with shrinking of the frontal and temporal anterior lobes of the brain.

Gerstmann's Syndrome: Gerstmann's syndrome is a cognitive impairment that results from damage to a specific area of the brain -- the left parietal lobe in the region of the angular gyrus.

Gerstmann-Straussler-Scheinker Disease: Gerstmann-Straussler-Scheinker disease (GSS) is an extremely rare, neurodegenerative brain disorder. In the early stages, patients may experience varying levels of ataxia (lack of muscle coordination), including clumsiness, unsteadiness, and difficulty walking.

Globoid Cell Leukodystrophy: Globoid cell leukodystrophy is a rare, inherited degenerative disorder of the central and peripheral nervous systems. It is characterized by the presence of globoid cells (cells that have more than one nucleus), the breakdown of the nerve's protective myelin coating, and destruction of brain cells.

Hallervorden-Spatz Disease: Hallervorden-Spatz disease is neurodegeneration with brain iron accumulation (NBIA) and is a rare, inherited, neurological movement disorder characterized by progressive degeneration of the nervous system.

Hemicrania Continua: Hemicrania continua is a rare form of chronic headache marked by continuous pain on one side of the face that varies in severity.

Hemiplegia Alternia: Alternating hemiplegia is a rare neurological disorder that develops in childhood, most often before the child is 18 months old. The disorder is characterized by recurrent episodes of paralysis that involve one or both sides of the body, multiple limbs, or a single limb.

Hereditary Neuropathies: Hereditary neuropathies are a group of inherited disorders affecting the peripheral nervous system. The hereditary neuropathies are divided into four major subcategories: hereditary motor and sensory neuropathy, hereditary sensory neuropathy, hereditary motor neuropathy, and hereditary sensory and autonomic neuropathy.

Hereditary Spastic Paraplegia: Hereditary spastic paraplegia (HSP), also called familial spastic paraparesis (FSP), refers to a group of inherited disorders that are characterized by progressive weakness and spasticity (stiffness) of the legs. Early in the disease course, there may be mild gait difficulties and stiffness.

Hirayama Syndrome: Hirayama syndrome or monomelic amyotrophy (MMA) is characterized by progressive degeneration and loss of motor neurons, the nerve cells in the brain and spinal cord that are responsible for controlling voluntary muscles.

Holoprosencephaly: Holoprosencephaly is a disorder caused by the failure of the prosencephalon (the embryonic forebrain) to sufficiently divide into the double lobes of the cerebral hemispheres. The result is a single-lobed brain structure and severe skull and facial defects.

HTLV-1 Associated Myelopathy: HTLV-1 Associated Myelopathy has been used to describe a chronic and progressive disease of the nervous system that affects adults living in equatorial areas of the world and causes progressive weakness, stiff muscles, muscle spasms, sensory disturbance, and sphincter dysfunction.

Huntington's Disease: Huntington's disease (HD) results from genetically programmed degeneration of brain cells, called neurons, in certain areas of the brain. This degeneration causes uncontrolled movements, loss of intellectual faculties, and emotional disturbance.

Hydranencephaly: Hydranencephaly is a rare condition in which the brain's cerebral hemispheres are absent and replaced by sacs filled with cerebrospinal fluid.

Hydrocephalus: Hydrocephalus is a condition in which the primary characteristic is excessive accumulation of fluid in the brain.

Hydromyelia: Hydromyelia refers to an abnormal widening of the central canal of the spinal cord that creates a cavity in which cerebrospinal fluid (commonly known as spinal fluid) can accumulate.

Hypertonia: Hypertonia is a condition marked by an abnormal increase in muscle tension and a reduced ability of a muscle to stretch. It is caused by injury to motor pathways in the central nervous system, which carry information from the central nervous system to the muscles and control posture, muscle tone, and reflexes. Hypotonia may be detected by noting that the limbs can easily be displaced with little force, and that arm excursion during walking is increased

Hypoxia: Cerebral hypoxia refers to a condition in which there is a decrease of oxygen supply to the brain even though there is adequate blood flow.

Immune-Mediated Encephalomyelitis: Acute disseminated encephalomyelitis (ADEM) is characterized by a brief but intense attack of inflammation in the brain and spinal cord that damages myelin – the protective covering of nerve fibers.

Infantile Neuroaxonal Dystrophy: Infantile neuroaxonal dystrophy (INAD) is a rare inherited neurological disorder. It affects axons, the part of a nerve cell that carries messages from the brain to other parts of the body, and causes progressive loss of vision, muscular control, and mental skills.

Infantile Refsum Disease: Infantile Refsum disease (IRD) is one of a small group of genetic diseases called peroxisome biogenesis disorders (PBD), which are part of a larger group of diseases called the leukodystrophies. These are inherited conditions that damage the white matter of the brain and affect motor movements.

Intracranial Cysts: Intracranial cysts are cerebrospinal fluid-filled sacs that are located between the brain or spinal cord and the arachnoid membrane, one of the three membranes that cover the brain and spinal cord.

Isaac's Syndrome: Isaac's syndrome (also known as neuromyotonia, Isaac-Mertens syndrome, continuous muscle fiber activity syndrome, and quantal squander syndrome) is a rare neuromuscular disorder caused by hyperexcitability and continuous firing of the peripheral nerve axons that activate muscle fibers.

Joubert Syndrome: Joubert syndrome is a rare brain malformation characterized by the absence or underdevelopment of the *cerebellar vermis* - an area of the brain that controls balance and coordination. The most common features of Joubert syndrome in infants include abnormally rapid breathing (hyperpnea), decreased muscle tone (hypotonia), jerky eye movements (oculomotor apraxia), mental retardation, and the inability to coordinate voluntary muscle movements (ataxia).

Kinsbourne Syndrome: Kinsbourne syndrome is a rare neurological disorder characterized by an unsteady, trembling gait, myoclonus (brief, shock-like muscle spasms), and opsoclonus (irregular, rapid eye movements).

Kleine-Levin Syndrome: Kleine-Levin syndrome is a rare disorder that primarily affects adolescent males (approximately 70 percent of those with Kleine-Levin syndrome are male). It is characterized by recurring but reversible periods of excessive sleep (up to 20 hours per day).

Klüver-Bucy Syndrome: Klüver-Bucy syndrome is a rare behavioral impairment that is associated with damage to both of the anterior temporal lobes of the brain. It causes individuals to put objects in their mouths and engage in inappropriate sexual behavior. Other symptoms may include visual agnosia (inability to visually recognize objects), loss of normal fear and anger responses, memory loss, distractibility, seizures, and dementia.

Krabbe Disease: Krabbe disease is a rare, inherited degenerative disorder of the central and peripheral nervous systems. It is characterized by the presence of globoid cells (cells that have more than one nucleus), the breakdown of the nerve's protective myelin coating, and destruction of brain cells.

Lambert-Eaton Myasthenic Syndrome: Lambert-Eaton myasthenic syndrome (LEMS) is a disorder of the neuromuscular junction-the site where nerve cells meet muscle cells and help activate the muscles.

Landau-Kleffner Syndrome: Landau-Kleffner syndrome (LKS) is a rare, childhood neurological disorder characterized by the sudden or gradual development of aphasia (the inability to understand or express language) and an abnormal electro-encephalogram (EEG). LKS affects the parts of the brain that control comprehension and speech.

Lateral Medullary Syndrome: Lateral Medullary Syndrome or Wallenberg's syndrome is a neurological condition caused by a stroke in the vertebral or posterior inferior cerebellar artery of the brain stem. Symptoms include difficulties with swallowing, hoarseness, dizziness, nausea and vomiting, rapid involuntary movements of the eyes (nystagmus), and problems with balance and gait coordination.

Learning Disabilities: Learning disabilities are disorders that affect the ability to understand or use spoken or written language, do mathematical calculations, coordinate movements, or direct attention.

Lennox-Gastaut Syndrome: Lennox-Gastaut syndrome is a severe form of epilepsy.

Leukodystrophy: Leukodystrophy refers to progressive degeneration of the white matter of the brain due to imperfect growth or development of the myelin sheath, the fatty covering that acts as an insulator around nerve fiber.

Lewy Body Dementia: Lewy Body dementia) is one of the most common types of progressive dementia. The central feature of DLB is progressive cognitive decline, combined with three additional defining features: (1) pronounced "fluctuations" in alertness and attention, such as frequent drowsiness, lethargy, lengthy periods of time spent staring into space, or disorganized speech; (2)

recurrent visual hallucinations, and (3) parkinsonian motor symptoms, such as rigidity and the loss of spontaneous movement.

Lissencephaly: Lissencephaly, which literally means "smooth brain," is a rare, gene-linked brain malformation characterized by the absence of normal convolutions (folds) in the cerebral cortex and an abnormally small head (microcephaly).

Locked-in Syndrome: Locked-in syndrome is a rare neurological disorder characterized by complete paralysis of voluntary muscles in all parts of the body except for those that control eye movement.

Lou Gehrig's Disease: Lou Gehrig's disease, sometimes called Amyotrophic lateral sclerosis (ALS), is a rapidly progressive, invariably fatal neurological disease that attacks the nerve cells *(neurons)* responsible for controlling voluntary muscles.

Lupus: Lupus (also called *systemic lupus erythematosus*) is a disorder of the immune system. Lupus can cause other neurological disorders, such as mild cognitive dysfunction, organic brain syndrome, peripheral neuropathies, sensory neuropathy, psychological problems (including personality changes, paranoia, mania, and schizophrenia), seizures, transverse myelitis, and paralysis and stroke.

Megalencephaly: Megalencephaly, also called macrencephaly, is a condition in which an infant or child has an abnormally large, heavy, and usually malfunctioning brain.

Meningitis: Meningitis and encephalitis are inflammatory diseases of the membranes that surround the brain and spinal cord and are caused by bacterial or viral infections. Viral meningitis is sometimes called aseptic meningitis to indicate it is not the result of bacterial infection and cannot be treated with antibiotics. Symptoms of encephalitis include sudden fever, headache, vomiting, heightened sensitivity to light, stiff neck and back, confusion and impaired judgment, drowsiness, weak muscles, a clumsy and unsteady gait, and irritability.

Microcephaly: Microcephaly is a medical condition in which the circumference of the head is smaller than normal because the brain has not developed properly or has stopped growing.

Migraine: Migraine headache is often described as an intense pulsing or throbbing pain in one area of the head. It is often accompanied by extreme sensitivity to light and sound, nausea, and vomiting.

Mini-Strokes: A transient ischemic attack (TIA) is a transient stroke that lasts only a few minutes. It occurs when the blood supply to part of the brain is briefly interrupted.

Monomelic Amyotrophy: Monomelic amyotrophy (MMA) is characterized by progressive degeneration and loss of motor neurons, the nerve cells in the brain and spinal cord that are responsible for controlling voluntary muscles.

Motor Neuron Disease: The motor neuron diseases (MNDs) are a group of progressive neurological disorders that destroy cells that control essential muscle activity such as speaking, walking, breathing, and swallowing.

Moyamoya Disease: Moyamoya disease is a rare, progressive cerebrovascular disorder caused by blocked arteries at the base of the brain in an area called the basal ganglia.

Multifocal Motor Neuropathy: Multifocal motor neuropathy is a progressive muscle disorder characterized by muscle weakness in the hands, with differences from one side of the body to the other in the specific muscles involved.

Multi-Infarct Dementia: Multi-infarct dementia (MID) is a common cause of memory loss in the elderly. MID is caused by multiple strokes (disruption of blood flow to the brain).

Multiple Sclerosis: An unpredictable disease of the central nervous system, multiple sclerosis (MS) can range from relatively benign to somewhat disabling to devastating, as communication between the brain and other parts of the body is disrupted.

Myasthenia-Congenita: All forms of myasthenia are due to problems in the communication between nerve cells and muscles, which involve the activities of neurotransmitters.

Myelinoclastic Diffuse Sclerosis: It is also called Schilder's disease which is not the same as Addison-Schilder disease (adrenoleukodystrophy). Schilder's disease is a rare progressive demyelinating disorder which usually begins in childhood. Symptoms may include dementia, aphasia, seizures, personality changes, poor attention, tremors, balance instability, incontinence, muscle weakness, headache, vomiting, and vision and speech impairment.

Myopathy – Congenital: A myopathy is a disorder of the muscles that usually results in weakness. Congenital myopathy refers to a group of muscle disorders that appear at birth or in infancy. Typically, an infant with a congenital myopathy will be "floppy," have difficulty breathing or feeding, and will lag behind other babies in meeting normal developmental milestones such as turning over or sitting up.

Narcolepsy: Narcolepsy is a chronic neurological disorder caused by the brain's inability to regulate sleep-wake cycles normally.

Neuroacanthocytosis: Neuroacanthocytosis refers to a group of genetic conditions that are characterized by movement disorders and acanthocytosis (abnormal, spiculated red blood cells). Four syndromes are classified as neuroacanthocytosis: Chorea-acanthocytosis, McLeod syndrome, Huntington's disease-like 2 (HDL2), and panthothenate kinase-associated neurodegeneration (PKAN).

Neurofibromatosis: Neurofibromatoses are genetic disorders of the nervous system that primarily affect the development and growth of neural (nerve) cell tissues. These disorders cause tumors to grow on nerves and produce other abnormalities such as skin changes and bone deformities.

Neuroleptic malignant syndrome: Neuroleptic malignant syndrome is a life-threatening, neurological disorder most often caused by an adverse reaction to neuroleptic or antipsychotic drugs. Symptoms include high fever, sweating, unstable blood pressure, stupor, muscular rigidity, and autonomic dysfunction.

Neuromyotonia: Isaac's syndrome (also known as neuromyotonia, Isaac-Mertens syndrome, continuous muscle fiber activity syndrome, and quantal squander syndrome) is a rare neuromuscular disorder caused by hyperexcitability and continuous firing of the peripheral nerve axons that activate muscle fibers.

Neuronal Ceroid Lipofuscinosis: Batten disease is a fatal, inherited disorder of the nervous system that begins in childhood. In some cases, the early signs are subtle, taking the form of personality and behavior changes, slow learning, clumsiness, or stumbling.

Neurosarcoidosis: Neurosarcoidosis is a manifestation of sarcoidosis in the nervous system. Sarcoidosis is a chronic inflammatory disorder that typically occurs in adults between 20 and 40 years of age and primarily affects the lungs, but can also impact almost every other organ and system in the body.

Neurosyphilis: Neurosyphilis is a disease of the coverings of the brain, the brain itself, or the spinal cord. It can occur in people with syphilis, especially if they are left untreated. Neurosyphilis is different from syphilis because it affects the nervous system, while syphilis is a sexually transmitted disease with different signs and symptoms.

Nevus Cavernosus: A cerebral cavernous malformation (CCM) is a collection of small blood vessels (capillaries) in the central nervous system (CNS) that is enlarged and irregular in structure.

Neuralgia –Occipital: Neuralgia –Occipital is a distinct type of headache characterized by piercing, throbbing, or electric-shock-like chronic pain in the upper neck, back of the head, and behind the ears, usually on one side of the head.

Ohtahara Syndrome: Ohtahara syndrome is a neurological disorder characterized by seizures.

Olivopontocerebellar Atrophy: Olivopontocerebellar atrophy (OPCA) is a term that describes the degeneration of neurons in specific areas of the brain – the cerebellum, pons, and inferior olives.

Opsoclonus myoclonus: Opsoclonus myoclonus is a rare neurological disorder characterized by an unsteady, trembling gait, myoclonus (brief, shock-like muscle spasms), and opsoclonus (irregular, rapid eye movements).

O'Sullivan-McLeod Syndrome: O'Sullivan-McLeod syndrome is also called monomelic amyotrophy (MMA) which is characterized by progressive degeneration and loss of motor neurons, the nerve cells in the brain and spinal cord that are responsible for controlling voluntary muscles.

Paresthesia: Paresthesia refers to a burning or prickling sensation that is usually felt in the hands, arms, legs, or feet, but can also occur in other parts of the body. The sensation, which happens without warning, is usually painless and described as tingling or numbness, skin crawling, or itching.

Parkinson's Disease: Parkinson's disease (PD) belongs to a group of conditions called motor system disorders, which are the result of the loss of dopamine-producing brain cells.

Parry-Romberg syndrome: Parry-Romberg syndrome is a rare disorder characterized by slowly progressive deterioration (atrophy) of the skin and soft tissues of half of the face (hemifacial atrophy), usually the left side.

Pelizaeus-Merzbacher disease: Pelizaeus-Merzbacher disease (PMD) is a rare, progressive, degenerative central nervous system disorder in which coordination, motor abilities, and intellectual function deteriorate. The disease is one of a group of gene-linked disorders known as the leukodystrophies, which affect growth of the myelin sheath -- the fatty covering that wraps around and protects nerve fibers in the brain.

Pena Shokeir II Syndrome: Cerebro-oculo-facio-skeletal syndrome (COFS) is a pediatric, genetic, degenerative disorder that involves the brain and the spinal cord. It is characterized by craniofacial and skeletal abnormalities, severely reduced muscle tone, and impairment of reflexes.

Perineural Cysts: Perineural cysts are sacs filled with cerebrospinal fluid that most often affect nerve roots in the sacrum, the group of bones at the base of the spine.

Periventricular Leukomalacia: Periventricular leukomalacia (PVL) is characterized by the death of the white matter of the brain due to softening of the brain tissue. It can affect fetuses or newborns; premature babies are at the greatest risk of the disorder.

Pick's Disease: Pick's disease is also known as frontotemporal dementia (FTD) describes a clinical syndrome associated with shrinking of the frontal and temporal anterior lobes of the brain.

Porencephaly: Porencephaly is an extremely rare disorder of the central nervous system in which a cyst or cavity filled with cerebrospinal fluid develops in the brain.

Postinfectious Encephalomyelitis: Postinfectious Encephalomyelitis is characterized by a brief but intense attack of inflammation in the brain and spinal cord that damages myelin – the protective covering of nerve fibers.

Primary Dentatum Atrophy: Primary dentatum atrophy refers to a collection of rare, degenerative, neurological disorders characterized by epilepsy, cognitive impairment, myoclonus, and progressive ataxia.

Prion Diseases: Prion diseases is transmissible spongiform encephalopathies (TSEs), also known as prion diseases, are a group of rare degenerative brain disorders characterized by tiny holes that give the brain a "spongy" appearance.

Pseudotumor Cerebri: Pseudotumor cerebri literally means "false brain tumor." It is likely due to high pressure within the skull caused by the buildup or poor absorption of cerebrospinal fluid (CSF). The disorder is most common in women between the ages of 20 and 50.

Ramsay Hunt Syndrome I: Ramsay Hunt syndrome I refers to a collection of rare, degenerative, neurological disorders characterized by epilepsy, cognitive impairment, myoclonus, and progressive ataxia.

Rasmussen's Encephalitis: Rasmussen's encephalitis is a rare, chronic inflammatory disease that usually affects only one hemisphere of the brain. It occurs in children under the age of 10 (and more rarely in adolescents and adults), and is characterized by frequent and severe seizures, loss of motor skills and speech, paralysis on one side of the body (hemiparesis), inflammation of the brain (encephalitis), and mental deterioration.

Refsum Disease: Refsum disease (ARD) is one of a group of genetic diseases called leukodystrophies, which damage the white matter of the brain and affect motor movements.

Reye's Syndrome: Reye's syndrome (RS) is primarily a children's disease, although it can occur at any age. It affects all organs of the body but is most harmful to the brain and the liver--causing an acute increase of pressure within the brain and, often, massive accumulations of fat in the liver and other organs.

Sacral Nerve Root Cysts: Sacral nerve root cysts are sacs filled with cerebrospinal fluid that most often affect nerve roots in the sacrum, the group of bones at the base of the spine.

Schizencephaly: Schizencephaly is an extremely rare developmental birth defect characterized by abnormal slits, or clefts, in the cerebral hemispheres of the brain.

Seitelberger Disease: Seitelberger disease is a rare inherited neurological disorder. It affects axons, the part of a nerve cell that carries messages from the brain to other parts of the body, and causes progressive loss of vision, muscular control, and mental skills.

Shy-Drager Syndrome: Shy-Drager syndrome is the current classification for a neurological disorder that was once called Shy-Drager syndrome. A progressive disorder of the central and autonomic nervous systems, it is characterized by orthostatic hypotension (an excessive drop in blood pressure when standing up) which causes dizziness or fainting.

Sleep apnea: Sleep apnea is a common sleep disorder characterized by brief interruptions of breathing during sleep. These episodes usually last 10 seconds or more and occur repeatedly throughout the night.

Spasticity: Spasticity is a condition in which certain muscles are continuously contracted. This contraction causes stiffness or tightness of the muscles and may interfere with movement, speech, and manner of walking.

Spina Bifida: Spina bifida (SB) is a neural tube defect (a disorder involving incomplete development of the brain, spinal cord, and/or their protective coverings) caused by the failure of the fetus's spine to close properly during the first month of pregnancy.

Steele-Richardson-Olszewski Syndrome: Steele-Richardson-Olszewski is a rare brain disorder that causes serious and permanent problems with control of gait and balance. The most obvious sign of the disease is an inability to aim the eyes properly, which occurs because of lesions in the area of the brain that coordinates eye movements.

Striatonigral degeneration: Striatonigral Degeneration is a neurological disorder caused by a disruption in the connection between two areas of the brain-the striatum and the substantia nigra.

Syncope: Syncope is the temporary loss of consciousness due to a sudden decline in blood flow to the brain. It may be caused by an irregular cardiac rate or rhythm or by changes of blood volume or distribution.

Syringohydromyelia: Syringohydromyelia refers to an abnormal widening of the central canal of the spinal cord that creates a cavity in which cerebrospinal fluid (commonly known as spinal fluid) can accumulate. As spinal fluid builds up, it may put abnormal pressure on the spinal cord and damage nerve cells and their connections.

Syringomyelia: Syringomyelia is a disorder in which a cyst forms within the spinal cord. This cyst, called a syrinx, expands and elongates over time, destroying the center of the spinal cord. Since the spinal cord connects the brain to nerves in the extremities, this damage results in pain, weakness, and stiffness in the back, shoulders, arms, or legs.

Tabes Dorsalis: Tabes dorsalis is a slow degeneration of the nerve cells and nerve fibers that carry sensory information to the brain.

Tarlov Cysts: Tarlov cysts are sacs filled with cerebrospinal fluid that most often affect nerve roots in the sacrum, the group of bones at the base of the spine.

Tethered Spinal: Tethered spinal cord syndrome is a neurological disorder caused by tissue attachments that limit the movement of the spinal cord within the spinal column.

Tourette syndrome: Tourette syndrome (TS) is a neurological disorder characterized by repetitive, stereotyped, involuntary movements and vocalizations called tics.

Tropical Spastic Paraparesis: Tropical spastic paraparesis (TSP) has been used to describe a chronic and progressive disease of the nervous system that affects adults living in equatorial areas of the world and causes progressive weakness, stiff muscles, muscle spasms, sensory disturbance, and sphincter dysfunction.

Tuberous sclerosis: Tuberous sclerosis (TSC) is a rare genetic disease that causes benign tumors to grow in the brain and on other vital organs such as the kidneys, heart, eyes, lungs, and skin.

Vascular Erectile Tumor: Vascular erectile tumor is a collection of small blood vessels (capillaries) in the central nervous system (CNS) that is enlarged and irregular in structure.

Von Economo's Disease: Von Economo's disease is a disease characterized by high fever, headache, double vision, delayed physical and mental response, and lethargy. In acute cases, patients may enter coma.

Von Hippel-Lindau Disease: Von Hippel-Lindau disease (VHL) is a rare, genetic multi-system disorder characterized by the abnormal growth of tumors in certain parts of the body (angiomatosis).

Wallenberg's Syndrome: Wallenberg's syndrome is a neurological condition caused by a stroke in the vertebral or posterior inferior cerebellar artery of the brain stem. Symptoms include difficulties with swallowing, hoarseness, dizziness, nausea and vomiting, rapid involuntary movements of the eyes (nystagmus), and problems with balance and gait coordination.

Wernicke's Encephalopathy: Wernicke's encephalopathy is a degenerative brain disorder caused by the lack of thiamine (vitamin B1). It may result from alcohol abuse, dietary deficiencies, prolonged vomiting, eating disorders, or the effects of chemotherapy. Symptoms include mental confusion, vision impairment, stupor, coma, hypothermia, hypotension, and ataxia.

Williams Syndrome: Williams syndrome (WS) is a rare genetic disorder characterized by mild to moderate mental retardation or learning difficulties, a distinctive facial appearance, and a unique personality that combines over-friendliness and high levels of empathy with anxiety. The most significant medical problem associated with WS is cardiovascular disease caused by narrowed arteries.

Wilson's Disease: Wilson's disease (WD) is a rare inherited disorder in which excessive amounts of copper accumulate in the body. The buildup of copper leads to damage in the kidneys, brain, and eyes.

X-Linked Spinal and Bulbar Muscular Atrophy: X-Linked Spinal and Bulbar Muscular Atrophy (or Kennedy's disease) is an inherited motor neuron disease that affects males. It is one of a group of disorders called *spinal muscular atrophy* (SMA). Early symptoms include tremor of the outstretched hands, muscle cramps with exertion, and fasciculations (fleeting muscle twitches visible under the skin). Eventually, individuals develop limb weakness which usually begins in the pelvic or shoulder regions.